"This is a highly innovative book, which explores the n
family travel, focussing particularly on how it influenc
'future-making projects' of both families and individuals. Drawing on data
from parents and young people, and including the perspectives of those who
reject travel (for environmental reasons) as well as those who travel a lot, the
book engages with important debates that cut across several disciplines – including the extent to which such practices contribute to social differentiation,
how mobility is conceptualised, and the role of travel within broader understandings of parenting. It is also international in its orientation, and will thus
be of significant interest to researchers in many different national contexts."
—**Rachel Brooks** *FAcSS, Professor of Sociology,*
University of Surrey, UK

"This is a thought-provoking book on how mobility shapes individuals. By
zooming into the experiences of family travel, the book offers engaging accounts of the links between the ideas of mobility and social class. Highly rec-
ommended for sociologists of education and social scientists more broadly."
—**Maia Chankseliani**, *Associate Professor*
of Comparative and International Education,
University of Oxford, UK

"Maxwell, Yemini and Bach offer a rigorous and thoughtful journey into
some of the uncharted aspects of mobility, by exploring family travel and its
nuanced links with parenting, family-making practices, strategies of capital
accumulation and class differentiations."
—**Jason Beech**, *Senior Lecturer in Education Policy,*
Monash University, Australia

Nurturing Mobilities

Nurturing Mobilities employs new empirical material and an innovative theoretical framing to bring new clarity to why families travel today – and what happens when they do. The authors argue that an imperative to 'think with mobility' and to 'aspire to be mobile' shapes identities, futures and family practices.

Drawing on data that examines family travel practices – typically short-term trips – across the working-, middle-, and globally mobile middle-classes, *Nurturing Mobilities* describes how families travel, why they travel, and the role young family members play in curating family travel. Vitally, it examines the two biggest contemporary issues in global mobility: COVID-19 and climate change. How has COVID-19 changed travel motivations in a world beset by lockdowns and diminished finances? How are concerns around climate change, and engagements with global citizenship education, changing family travel practices?

Nurturing Mobilities illuminates new ways in which social class divergence is forged through movements across borders. The authors' theoretically inter-disciplinary approach delivers a full analysis of the apparently divergent processes that differentiate family travel along social class lines, yet also allow travel to play a core role in social mobility. This book is a vital resource for scholars and students studying mobility, globalisation, social class, and climate change engagement.

Claire Maxwell is a professor of sociology at the University of Copenhagen, Denmark. Her research has focused on the ways the internationalisation of education has shaped education systems and how elite forms of provision are being developed and embedded around the world. A second focus has been on the lives of globally mobile professionals and their families, examining school choice, identities and family practices.

Miri Yemini is a comparative education scholar at Tel Aviv University, Israel, with interests in internationalisation of education in schools and higher education, global citizenship education, and education in conflict-ridden societies, with a particular focus on the role of mobility in educational experiences. Dr. Yemini is an active member of CIES, CESE and BAICE and she is a President Elect for the Israeli Comparative Education Society.

Katrine Mygind Bach holds a Masters in Global Development and a Bachelor degree in Sociology from the University of Copenhagen, Denmark. Previously, she spent several years in India working in the field of rural development. Her research focuses on the intersection between social and environmental sustainability, and the role of climate governance in these processes.

Networked Urban Mobilities Series

Editors: Sven Kesselring, Nürtingen-Geislingen University, Germany and Malene Freudendal-Pedersen, Roskilde University, Denmark

The Networked Urban Mobilities series resulted from the Cosmobilities Network of mobility research and the Taylor & Francis journal, 'Applied Mobilities.' This three volume set, ideal for mobilities researchers and practitioners, explores a broad number of topics including planning, architecture, geography and urban design.

Exploring Networked Urban Mobilities
Theories, Concepts, Ideas
Malene Freudendal-Pedersen and Sven Kesselring

Experiencing Networked Urban Mobilities
Sites, Methods, Practices
Malene Freudndal-Pedersen, Katrine Hartmann-Petersen and Emmy Laura Perez Fjalland

Envisioning Networked Urban Mobilities
Art, Creativity, Performance
Aslak Aamot Kjaerulff, Sven Kesselring, Peter Peters and Kevin Hannam

Sharing Mobilities
New Perspectives for the Mobile Risk Society
Edited by Malene Freudendal-Pedersen, Sven Kesselring and Dennis Zuev

Nurturing Mobilities
Family Travel in the 21st Century
Claire Maxwell, Miri Yemini and Katrine Mygind Bach

For more information about this series, please visit: https://www.routledge.com/Networked-Urban-Mobilities-Series/book-series/NUM

Nurturing Mobilities

Family Travel in the 21st Century

**Claire Maxwell,
Miri Yemini and
Katrine Mygind Bach**

Routledge
Taylor & Francis Group

NEW YORK AND LONDON

First published 2022
by Routledge
605 Third Avenue, New York, NY 10158

and by Routledge
2 Park Square, Milton Park, Abingdon, Oxon, OX14 4RN

Routledge is an imprint of the Taylor & Francis Group, an informa business

Library of Congress Cataloging-in-Publication Data
A catalog record for this title has been requested

ISBN: 9780367520939 (hbk)
ISBN: 9781032114811 (pbk)
ISBN: 9781003056430 (ebk)

DOI: 10.4324/9781003056430

Typeset in Bembo
by codeMantra

Contents

Acknowledgements

Thank-you to all the participants who gave their time to become involved in this exploration of family holiday travel, and to the three University of Copenhagen BA sociology students (Maluhs, Olivia, and Camilla) who contributed to the development of two central chapters. The research and development of this book was partially supported by a grant from the Israeli Science Foundation 222/20.

1 Introduction

This book seeks to establish a new agenda in the scholarship of family travel, with a focus on short-term travel. Such a focus is necessary as so many families, not only the most highly resourced, are now mobile. With cheaper air travel, an increasing imperative to 'see the world', a commitment by schools to ensure their students see themselves situated in a broader transnational space through the internationalisation of education, religious pilgrimages, and a desire to maintain links to families spread all over the world, more and more families are mobile. Yet no book exists to date that examines the various aspects of short-term family travel, and none does this through a sociological lens. Therefore, this book examines the ways parents, young adults, and children from various socio-economic and spatial backgrounds engage with the act of travelling, exploring whether and how this shapes understandings of their positionality, not only in more local but also in more global spaces.

Our data has focussed on examining forms of short-term travel, reasons behind it, how destinations are decided on, the experiences had while travelling, and parents and young people's reflections about the act of travel and its longer-term consequences. In being mobile across national border, we have considered what resources make this possible. We have focussed on whether and how short-term travel shapes individual and family identities, how it becomes integrated into parents' and young people's own 'future making' projects, and how international travel opens up engagements with the 'Other', as well as reflecting back our belonging to the 'home' nation. In light of growing concerns over the consequences of (air) travel on climate change, we have extended our focus through examining the views of people who abstain from travel or try to practice a 'no fly' form of travel. Also significant is our inclusion of families from the so-called global middle class – a growing faction within the middle classes – for whom mobility is more integrated into their lives. We have examined whether and how engagements with short-term leisure travel intersect with other forms of mobility for them, and whether or not these mobilities are different to those of their locally moored middle-class peers. Our findings lead us to argue that international travel illuminates new ways that social class differentiations are articulated, with interesting consequences for identity construction and potential for social

DOI: 10.4324/9781003056430-1

mobility. The book is primarily focussed on the narratives of parents and how they integrate travel practices into their education and family-making practices. However, we have also brought in the necessary perspectives of young people and young adults, to offer further insights, enabling us to examine this topic in a more multifaceted way.

We situate our book within a time when processes of globalisation continue to alter economic and political structures and relations. In many cases, mobility has been made easier and cheaper, but in other situations, the right to movement is becoming more restricted – both for leisure and migration (Harpaz, 2019) – in light of tightening immigration rules in some countries and the closure of borders during the time of a pandemic, as with COVID-19. Our work is firmly emplaced within the 'mobility turn' paradigm (Urry, 2007). In investigating closely small and larger differentiations within family travel practices across social class lines and factions, we understand short-term mobility as multidimensional and complicated by space and time. We are therefore able to illustrate, through our close study of parents and children, how new distinctions between social groups are created through family travel, and, in particular, orientations to travel (where, how, and why), which in turn help make sense of further social striations found across societies. Understanding the role of travel – a mechanism of discovery but also a form of capital accumulation in securing educational and social advantage – is therefore critical for the study of sociology.

Theoretical Framework

The premise underpinning our work is that mobility is a fundamental social issue (Easthope, 2009). The ease of physical and virtual mobility has been widely acknowledged as critical in transforming peoples' lives in the 21st century, leading to a re-shaping of relations within and between nation-states. Sheller and Urry (2006, p. 207) argue that today, more than ever, 'all the world seems to be on the move'. The nature of physical forms of mobility and the reasons behind these will vary, from people seeking asylum through those fleeing from war, environmental disasters, and political persecution to families who are migrating more overtly for economic and professional reasons, providing their expertise in different forms of employment – the service industry, government missions, and multinational corporations (Kaufmann et al., 2004). Furthermore, mobility has increased for all families through heightened expectations for, and the low cost of, international travel for holidays (Barker et al., 2009). Mobility is being conceptualised as central for human beings, while unjust access to various types of mobilities is being negotiated in legal, political, cultural, and social discourses (Sheller, 2018). Moreover, as suggested by Tanu (2018), being mobile has been largely conceptualised with regard to Western, white, and otherwise privileged populations, despite the increasing relevance of such practices for those outside the Global North as well.

In our book we are specifically interested in how those from differently located social classes, who nonetheless travel abroad for various reasons, articulate an understanding of travel as a form of mobility that becomes integrated into their everyday lives and shapes practices of cultivation in the present and future. We have aimed to broaden the concept of the tourist gaze to include other forms of mobilities which are increasingly common, such as visits to families scattered around the world as part of the diaspora, and mobilities for education. Such a focus allows us to consider how access to forms of physical mobility, especially that which takes them outside the nation-state, is understood as a resource and orientation to the world. To facilitate our research framing, we have combined several theoretical concepts, including Kaufmann et al.'s (2004) concept of motility, Lareau's (2003) work on family parenting practices, and Andreotti et al.'s (2015) understanding of global mindedness.

Flamm and Kaufmann (2006) theorise mobility as a form of capital; they argue mobility is not just about spatial movement, but rather about how such spatial movement is made possible through social position. Thus, according to Kaufmann et al. (2004), physical mobility not only affects an individual's or group's positioning within the social structure, but may also itself affect social structures more broadly. Kaufmann et al. (2004) therefore reconceptualise mobility as 'motility' in order to emphasise its association with a form of capital, and define motility as examining 'how entities access and appropriate the capacity for socio-spatial mobility' (Kaufmann et al., 2004, p. 750). Critically, Flamm and Kaufmann (2006) focus on both the potential (in terms of desire or perceived desirability) and actual capacity for mobility (access and skills) – alerting us to the importance of aspirations in shoring up practices. We draw on motility as a way of understanding family and parenting practices around the education and cultivation of children – to prepare them for physical and social mobility contemporaneously as well as in the future. More specifically, inspired by Kaufmann and colleagues (2004), we consider how our families make accessible different forms of physical mobility to their children, and the kinds of competencies they seek to instil in them as part of being physically mobile through travelling abroad. In this way we can consider how international travel is a resource capable of drawing social distinctions between groups, and is used by parents in conjunction with other forms of capital accumulation to secure their children's social position, and make accessible different spaces – at local, national, and transnational levels – or as a capital they seek to make available so as to promote their children's possibilities for social mobility (Maxwell & Yemini, 2019; Yemini & Maxwell, 2018; Yemini et al., 2019).

Supporting our key theoretical concept – that of motility – we furthermore lean on Lareau's (2003) concept of concerted cultivation which also draws on Bourdieu's work around capitals and habitus. Famously, Lareau argued that there was a classed cultural logic underpinning parental education strategies, where the middle and working classes appeared to practise their parenting differently. Yet, in one of the author's engagement with Lareau's theorisations, Vincent and Maxwell (2016) have argued that such a distinction between the differently resourced social groups in terms of intentions is perhaps less clear-cut. While access to resources may limit the extent, or the manner in which parents can pursue their desires to 'cultivate' their child, the responsibilisation and intensification of parenting are shown to affect all social groups and therefore their practices. In our current study we have therefore sought to examine closely the extent to which differently located social groups understand travel as a form of cultivation, and what resources they draw on to enact this. In doing so, we not only record who moves and who stays moored, but we also do more work in revealing the power structures that make sense of these movements, or as Doreen Massey (1994, p. 149) puts it – articulating the 'power geometry of space–time compression'.

Our focus on social class differentiations is largely set within a national frame of reference – where socio-cultural and -economic location is framed by the nation-state or at an even more local level. But travel outside the nation's borders also takes people into new cultural spaces, where they are either directly confronted with the 'Other' or have to actively choose to avoid too much of an immersion in new geographical, cultural, and social spaces. Thus, travel has the potential to create stronger bonds across differences or to further embed the largely post-colonial relations that still govern the power geometries across transnational spaces. To ensure we actively engaged with these questions, we explored how our participants articulated a disposition towards global mindedness when reflecting on their family travel experiences (Andreotti et al., 2015). Seeking to integrate such questions with broader concerns around social justice, we have also considered how family holiday travel can be a form of Global Citizenship Education and the extent to which it could precipitate new ways of thinking and doing travel that seek to minimise the environmental and imperial implications of travel by people in the Global North.

Although these were our three main organising concepts within which we framed our research and the initial analysis, as will be seen throughout the book, we have added in new concepts, when needed, to fully nuance the analysis and make sense of the data. In this way, our book therefore becomes a project not only to establish 'the field of family travel' within a sociological frame but also to offer a theoretical framework within which future research could be situated.

A Summary of the Subsequent Chapters

Chapter 2 starts off the book in earnest, by providing an overview of key research done on travel to date. There are three main foci emerging from the literature: focussing on differences and opportunities around travel practices across social groups, countries, and cultures; the meanings and experiences generated through short-term leisure travel; and the potential paradoxes thrown up by travel today. We show how the field of family travel has seen important academic developments both in terms of widening its object of study to encompass a broader representation of the different dimensions of travel, as well as addressing the nuances of how different social groups travel. Moreover, the evolution of the field of travel has impacted the forms of data and artefacts used to research patterns of short-term and leisure travel, which has effectively enriched the field and allowed for deeper engagements with the meanings of travel for people. The literature represented in this chapter is used as a stepping stone to introduce the further analyses of the role of family travel in the book.

Chapter 3 introduces in greater detail our theoretical framework of the book: Kaufmann's concept of motility, Andreotti et al.'s typology of global mindedness, and Lareau's work on families who occupy various social positionings and their parenting strategies.

Chapter 4 present our first, broad findings examining how parents talk about and understand travel. To do so, we draw on two sets of qualitative data, including in-depth research interviews with Danish parents and open-ended survey answers of parents residing in multiple countries across the Global North. We outline how parents talk about the purposes of travel, what elements go into ensuring the 'right' kind of family travel, as well as their views on the kinds of orientations and skills acquired while travelling and how these might be converted for use elsewhere. We show that these three considerations intersect in shaping an understanding of family travel for parents. First, children are central in shaping the practices around family travel, suggesting that travel is used as an element of parents' child-rearing strategies. This is shown by the way children's needs influence decisions about where to go and what to do, through the desire to create time and space for enriching family intimacy, and through a concern to improve their future, long-term opportunities. Second, 'going away' to foreign destinations is considered an integral part of being able to acquire the necessary space, but also skills and experiences that cannot be accessed 'at home' for family travel to be successful. Lastly, travel experiences in the present and near future are generally perceived to influence the family members' possibilities for developments around education, work, and mobility in the future, thereby reinforcing the importance of travel. While providing an initial investigation into how we might begin to think about family travel from a sociological perspective, this chapter also opens up the questions that are explored in more depth throughout the rest of the book.

In Chapter 5 we combine Kaufmann's notion of motility and Andreotti and colleagues' development of global mindedness dispositions to argue that Israeli families use family travel to practice cosmopolitan nationalism. This allows both working- and middle-class families to acquire some experience of the 'Other' through travel which will extend their aspirations and the possibilities for the transmission of this form of cosmopolitan encounters into cultural capital. Critically, however, travelling abroad is used to extend and embed their children's attachment to Israel, their 'home' nation. In this way we argue that family travel abroad is used as a practice of cosmopolitan nationalism. Yet differences in social location do matter in the experiences and outcomes of family travel. Working-class families, who are in one way or another marginalised in Israeli society due to their immigrant background, are able to develop more 'empathetic' and 'visiting' dispositions of global mindedness through their connections to families located around the world as part of the Jewish diaspora. Meanwhile, the middle-class families, who travel more frequently and to a larger number of destinations, have a less meaningful interaction with the 'Other' through travel, remaining at the level of the 'tourist', which is often enough to be able to draw on these cosmopolitan experiences instrumentally as part of extending their education and employment future opportunities, but simultaneously work to embed their nationalist identity.

Global middle class (GMC) families provide the focus for Chapter 6. Drawing on Kaufmann and colleagues (2004), we show how these families use mobility to cultivate motility as a resource for their children to draw on in the present and future (i.e. mobility becomes a form of capital). While mobility is similarly important for families from the working and middle classes, the GMC families integrate mobility into their family routines, thus constituting it as a 'natural' practice. In so doing, mobility (in the form of family travel in this case) is activated as an accessible resource and becomes a prominent class signifier. In this chapter, we show how access to frequent and culturally attuned travel allows these families to envisage a future untethered to the nation-state, and to cultivate identities in their children that should allow them to have a wide range of choices about where to live and work in the future.

While the majority of materials we collected for this project focussed on parents, in Chapter 7 we examine the experiences and views on travel among a group of young people. Seven focus group discussions (of between 2 and 6 students) were conducted with 24 school students in the ages of 14–16 years, at a Danish public school. Using Kaufmann et al.'s (2004) framework on motility we explore how young people from different socio-economic locations experience family travel, what aspirations they have for their futures, and how this is linked to particular resources around, and access to, mobility. We find that young people are aware of the different kinds of resources and capabilities needed in order to access different forms of mobility, and the benefits one can derive from travel. Furthermore, we find that young people incorporate aspirations for mobility into their anticipated futures. We also argue that since norms and desires around family travel are socially constructed and negotiated, travel practices become one measure through which social positioning is negotiated, hence emphasising the role mobility plays in creating mechanisms of inclusion and exclusion among young people.

In Chapter 8 we focus on two types of families – those who avoid travelling for environmental reasons and those who choose to live more nomadic styles of life by travelling with their children around the world and staying

weeks or months in certain locations. We analyse data gathered from interviews with parents, blogs, fora, and more traditional media, where families
presented and explained their choice and offered some detailed accounts on
how they perceive their travel practices and its effect on their children. The
choice to combine the analysis of these seemly highly contradictory travel
practices together is motivated by the high similarity found in many of their
testimonies, both in terms of the non-traditional travel choices made that aim
to challenge middle-class norms, and their desire to use these unique family
travel strategies for global citizenship education. While the book focusses on
short-term family travel, in this chapter we present two alternatives to this
type of travel, considering ways that disrupt hitherto accepted widespread
family travel practices and opening up possibilities for re-thinking travel and
our relationship to the world.

In our final substantive chapter engaging with our data – Chapter 9 –
we examine the reasons some people are questioning the use of air travel,
and draw on 13 in-depth interviews with Danes experiencing flight shame.
Translated from the Swedish word *flygskam*, flight shame was originally
coined by Swedish media in 2018 to denote a tendency among environmental
activists, in particular Fridays For Future, to frame the need to reduce greenhouse gas emissions as an issue of personal responsibility. In this chapter we
examine two questions. First, how do concerns about climate change shape
individual relationships to travel? And second, how are such relationships negotiated, transformed, and challenged when situated in the context of family
travel? The *carbon gaze* was identified as a common perspective among people
who experience flight shame, as a discursive disposition that shapes individual aspirations, experiences, and identities around travel. In the context of
family travel the *carbon gaze* represents a reflexive contestation to tacit and
routinised travel practices. Yet the participants share how they experience
significant strains between, on the one hand, abiding to their ambitions to
remain 'grounded' (i.e. not travel by plane), and, on the other, respecting
their inclination to be part of meaningful social collectives like the family.
This analysis speaks to the very real challenges involved with changing the
climate-damaging social practices of travel through individual consumption.

Chapter 10 concludes the book, offering a summary of our key findings
and reflecting on our key theoretical contributions. A detailed overview of
our research design and the various data collected is given in the Appendix.

We are very excited about our contribution being published in *Networked Urban Mobilities* book series, edited by Sven Kesselring and Malene
Freudendal-Pedersen. Family travel as a specific form of mobility has not, to
date, been featured in the edited volumes published. However, we believe
our examination of this particular and multifaceted form of mobility has clear
links to prominent themes found here: how mobilities shapes relations of
inequality, feelings of belonging, and connections between the local and the
global, to name a few. We look forward to future generative discussions with
colleagues in the Cosmobilities Network, an international research network,

scholars publishing in *Applied Mobilities* journal, and other researchers working in different spaces.

References

de Oliveira Andreotti, V., Biesta, G., & Ahenakew, C. (2015). Between the nation and the globe: Education for global mindedness in Finland. *Globalisation, Societies and Education, 13*(2), 246–259.

Barker, J., Kraftl, P., Horton, J., & Tucker, F. (2009). The road less travelled–new directions in children's and young people's mobility. *Mobilities, 4*(1), 1–10.

Easthope, H. (2009). Fixed identities in a mobile world? The relationship between mobility, place, and identity. *Identities: Global Studies in Culture and Power, 16*(1), 61–82.

Flamm, M., & Kaufmann, V. (2006). Operationalising the concept of motility: A qualitative study. *Mobilities, 1*(2), 167–189.

Harpaz, Y. (2019). *Citizenship 2.0: Dual nationality as a global asset.* Princeton University Press.

Kaufmann, V., Bergman, M. M., & Joye, D. (2004). Motility: Mobility as capital. *International Journal of Urban and Regional Research, 28*(4), 745–756.

Lareau, A. (2003). *Unequal childhoods: Class, race, and family life.* University of California Press.

Massey, D. (1994). *Space, place and gender.* University of Minnesota Press.

Maxwell, C., & Yemini, M. (2019). Modalities of cosmopolitanism and mobility: Parental education strategies of global, immigrant and local middle class Israelis. *Discourse, 40*(5), 616–632.

Sheller, M. (2018). *Mobility justice: The politics of movement in an age of extremes.* Verso Books.

Sheller, M., & Urry, J. (2006). The new mobilities paradigm. *Environment and Planning A, 38*(2), 207–226.

Tanu, D. (2018). *Growing up in transit: The politics of belonging at an international school.* Berghahn Books.

Urry, J. (2007). *Mobilities.* Polity Press.

Vincent, C., & Maxwell, C. (2016). Parenting priorities and pressures: Furthering understanding of 'concerted cultivation'. *Discourse: Studies in the Cultural Politics of Education, 37*(2), 269–281.

Yemini, M., & Maxwell, C. (2018). De-coupling or remaining closely coupled to 'home': Educational strategies around identity-making and advantage of Israeli global middle-class families in London. *British Journal of Sociology of Education, 39*(7), 1030–1044.

Yemini, M., Maxwell, C., & Mizrachi, M. A. (2019). How does mobility shape parental strategies–a case of the Israeli global middle class and their 'immobile' peers in Tel Aviv. *Globalisation, Societies and Education, 17*(3), 324–338.

2 Establishing the Field

Why We Travel and Why It Matters

Introduction to the Field of 'Travel'

Travel, especially beyond national borders, has been studied by a wide variety of academic disciplines, including history, archaeology, economics, geography, anthropology, and sociology (Aybek et al., 2015). Some economic historians have argued that the spatial movement of people across the globe started long before the academic world coupled mobility with the concept of globalisation (e.g. Wallerstein, 1974–1980). Yet most contemporary sociologists agree that the reconfiguration of social structures that occurred in the late 19th century, primarily, but not exclusively, as a result of the global expansion of the capitalist mode of production, alongside the invention of novel technologies in transportation and communication, paved the way for new possibilities of transgressing time and space in ways that had not been possible in any previous periods (Appadurai, 1996; Bauman, 1998; Castells, 2010; Giddens, 1991).

'Travel' as a concept has several dimensions and meanings, just like 'being at home' can have different modalities. Throughout the history of humankind, hunters and gatherers have travelled constantly, influenced by the seasons and environmental conditions, as some nomadic communities still do today. Geographical mobility was always been pursued by peoples, but its nature has arguably changed over the centuries, as the social order imposed various mobility structures for different people and classes (Harari, 2014). This meant that the 'rich and powerful' could travel, even have different homes in various localities, which they occupied depending on their needs and desires, while 'the poor' were sometimes forced to flee, due to environmental, economic, or political reasons. Yet family travel practices depend not only on economic resources but also on the particular cultural and social resources they have access to, as well as a number of more specific variables including age, number of children, health conditions of the family members, and political and historical contexts. In this book we focus on the concept of travel from a present-day viewpoint, investigating how differences across economic, social, and cultural resources may shape family practices of travel. In addition, we look at how ethical and environmental concerns challenge and reshape the way we conceive travel.

DOI: 10.4324/9781003056430-2

In the social sciences, different disciplines have applied their own conceptual logics and means of data collection to the field of travelling. Human migration studies, for example, have primarily investigated what Aybek et al. (2015) classified as long-term *re-location* of people moving from one geographical place to another, including labour migrants, refugees, and people living in diaspora (p. 5). Such studies have explored the role of human travelling in deconstructing and restructuring ethnic, cultural, and economic identities and relations within and across communities (e.g. Lucht, 2012; McKay, 2007), the changing political discourses on the social, cultural, and economic potentials and dilemmas of migration (e.g. de Haas, 2010; Gamlen, 2014), and the role of online communication platforms with regard to articulations and negotiations of identities across space (e.g. Horst, 2006; Nur-Muhammad et al., 2016). Another large body of literature within the social sciences has explored the everyday mobility flows across shorter distances, which by Aybek et al. (2015) has been defined as *circular mobility* (p. 5). These draw on cultural and urban fields of research and include studies such as Tjorring's (2016) mapping of gendered dwelling patterns within private Danish homes, or research exploring the different uses of public spaces, from urban transportation infrastructures (e.g. Grindlay et al., 2020) to public parks and benches (Larsen, 2004).

The subject of this book lies somewhere in the middle on the mobility spectrum that can be drawn from long-term re-location to everyday forms of circular-mobility, namely, short-term family travel. A significant body of literature which focusses on 'leisure time travel' attends to the management and promotion of tourism (e.g. Campo-Martínez et al., 2010; Prayag & Ryan, 2011). Often driven by commercial interests, quantitative tourism studies tend to apply causal models of push-and-pull factors of behaviour that treat subjects as consumers of space to understand what motivates people to travel (Aybek et al., 2015). Less concerned with tourism, in its economic and commercial sense, anthropological accounts on travel tend to focus on the variety of meanings that people attribute to different ways of travelling, as well as the group distinctions that occur when the traveller is confronted with 'the elsewhere' or 'the Other' (e.g. Salazar, 2018).

Across these various disciplinary contributions, there has been little more sociologically oriented analysis of travel for the purposes of leisure or holidaying. Furthermore, a specific focus of travel within families is almost entirely missing. Thus, our contribution in this book is to look at travel as a family project, for leisure/holidaying and other purposes, that takes them beyond national borders. Given that many more people travel for leisure beyond national borders today than ever before, it is important to understand the meaning of such spatial and geographical mobility in a globalised world, where it occurs for leisure and short-term purposes and where it occurs within the family unit. The book therefore examines ways parents and young people from various socio-economic and spatial backgrounds engage with the act of travelling, exploring whether and how this shapes understanding of their positionality in a broader world.

The evolution of the scholarship around the field of travel has impacted the forms of data and artefacts used to inform research. Contemporary researchers have criticised the field for being overly fixated on the spoken and written experiences of travelling, accusing scholars in the field of 'textocentrism' (Singhal & Rattine-Flaherty, 2006). To direct focus to the visual as well as performative practices of travelling, scholars have begun employing more visual data-collection methods including photography (e.g. Balomenou & Garrod, 2019), video (Simpson, 2011) and postcards (Milman, 2012). Research based on participant generated images (PGI) has offered a deeper insight into the embodied experiences of those who travel, and allowed a broader understanding of how experiences during travels are constructed and represented (e.g. Jutla, 2000; Markwell, 1997). Such studies have generated a more complex analytical understanding of the concept of travel and have in turn informed theoretical frameworks around explanations for mobility, such as Janta et al. (2015), who propose five different rationales for travelling: social relationships; the provision of care; affirmations of identities and roots; maintenance of territorial rights; and leisure tourism – suggesting that these often occur in a blurred and overlapping manner.

While earlier studies of family travel assumed that leisure time travel was taken up by heteronormative, white, middle-class families; in recent years this research has evolved and begun to acknowledge the diversity in families engaged in leisure travel, challenging the previous Euro-centric focus (Lucena et al., 2015). It has opened for the examination of issues such as ethnicity, religion, class, sexual identities, and personal preferences as well as national identities and the right to access certain places. The rise in environmental concerns related to air travel has also recently begun to shape the field of research related to family travel, as populations become increasingly aware of issues related to sustainability, and some have begun changing their travel behaviours accordingly. Most recently, of course, COVID-19 has fundamentally changed how we travel across borders, and potentially also how we perceive the risks and dangers associated with travelling.

In this chapter, we discuss relevant empirical scholarship that depicts what is known and what is yet to be explored in relation to the processes of mobility through family travel. Specifically, we focus on what we regard as three key aspects related to the phenomenon of family travel: first, the purposes and anticipated outcomes or benefits of travel, and whether these differ by family socio-economic and -cultural location; second, the meaning-generating processes of travelling, and how this has been and can be explored; and third, the paradox at the intersection of travel, cosmopolitanism as a form of cultural capital accumulation and environmental concerns around air travel, that have become salient points of debate in recent years. By tracing these three strands of literature we aim to give readers a preliminary understanding of the current state of knowledge in relation to each of them and create a foundation that we will build on in the chapters that follow.

Differential Travel Patterns across Cultures and Social Groups

According to Bauman (1998), rather than democratising the freedom to move, globalisation has laid bare a system of stratification, where 'access to global mobility' demarks new forms of privilege. To describe the modes of social stratification that globalisation has caused, Bauman draws on the image of the 'tourist' as someone guided by voluntary choice and lightness of movement, and contrasts it to the 'vagabond', understood as someone who moves only when and because they are forced to do so. Studies have confirmed that despite attempts to enable travel for the masses, the richest parts of the global population are still the most frequent flyers (Alcock et al., 2017). Due to the rise in disposable incomes among Global South middle classes along with the rise in availability of low-cost airlines, domestic and outbound tourism has been established as a phenomenon in Asian, African, and Latin and South American countries (Brown & Hall, 2008). Yet, despite decades of increases in both outbound and incoming tourism from and to the continents of the Global South, with the exception of Southeast Asia, they continue to have the least mobile populations in terms of leisure travel (UNWTO, 2018).

Empirical research has investigated the motivations behind international leisure travel. Using comparative quantitative methods, cross-cultural studies have suggested that tourism from the Middle East (Prayag & Hosany, 2014), Korea (Kim & Prideaux, 2005), and India (Prayag & Ryan, 2011) is driven by many of the same motivational desires as travellers from the Global North, namely, socialisation with family or friends residing in countries outside of the 'homeland'; experiences of novelty; accumulating prestige; and rest and relaxation. Such studies often build on existing categorisations of motivational factors and might therefore lack more inductive means of understanding how aspirations are shaped and put into action in different cultural and political contexts. Taking a more constructivist approach to the concept of motivation, Appadurai (1996) has argued that the increased flows of global goods and people also involve the circulation of images and ideas that work to reconfigure hopes, dreams, and aspirations of living a modern life. Travelling entails the mental creation of imaginations, hopes and expectations associated with geographical places. This idea is supported by a body of research investigating, among other things, the effects of tourism materials in constructing authentic and desirable places as tourist destinations through narratives and images (Morgan et al., 2007). Tourism advertisement, and particularly efforts of 'nation branding', have been shown to influence our desires concerning travel destinations (e.g. Butterfield et al., 1998; Fullerton et al., 2013). These examples demonstrate the dynamics of travel desires in a globalised world, their co-constructed nature (replete with influences of an industry seeking to sell spatial mobility for the purposes of holidaying) and raises the question of how uniform these practices of travel might be becoming.

Expanding Bauman's (1998, 2000) distinction between the tourists and the vagabonds, research shows that mobility stratification in a globalised world separates those who travel from those who stay physically immobile. However, it also draws a finer net of differentiation between diverse ways of travelling (Frändberg, 2009). A Bourdieusian sociological focus on the families' spatial mobility, would assume class differences shaping travel, suggesting working- and middle-class families would enact their social locations or attempt to enhance their capital resources through vacation practices or other types of family travel (Urry, 2002). Carlson et al. (2017), for example, examine class differences in families' child-rearing practices regarding transnational cultural capital and its acquisition. Although their research focusses on family perceptions of study abroad programmes, their interview data offers insights into how some families (particularly those most highly resourced) use family travel to groom and prepare their children for living abroad. Numerous studies using quantitative measures support such findings, arguing that class relations (e.g. Ying et al., 2016) – along with other social denominators like gender (e.g. Jucan & Jucan, 2013), and age (e.g. Capella & Greco, 1987) – go a long way in explaining differences in travel destination desires and experiences. The same characteristics also tend to determine which sources of information different people attribute their destination choices to, be it recommendations by family and friends, magazines and newspaper, or advertisement videos (Capella & Greco, 1987; Fong et al., 2017).

While marginal economic resources have historically limited holidaying practices among working class families, middle-class family travel tends to be child-centred, with children playing a major, if not final, role in the decision-making process around determining travel destinations, and activities undertaken during holidays (Tomić et al., 2018; Van der Eecken et al., 2019). Weenink (2008) has explored parental strategies and motives for cultivating cosmopolitanism as a form of capital (in its own right) for their children through travel. Cosmopolitanism is here understood as a moral and political ideal of world citizenship, that builds on the idea of global connectedness and an open-mindedness towards to the Other (Weenink, 2008, p. 1089). _Weenink's (2008) findings, from a mixed methods study, suggest that whereas parents' inclination to provide children with cosmopolitan capital is related to the parent's own cosmopolitan capital and their ambitions, it cannot be said to be related to their social class location. In this book, we aim to further examine potential variations in parental strategies among the travel practices of working-, middle-, and global middle–class families (see Chapters 5 and 6). We have also gathered insights directly from young people, thus starting to address the paucity of research in this area (Poria et al., 2005, 2014). Although we use class differences as one way of distinguishing the groups we explore, our analysis takes a layered approach that addresses the nuances and differences within each group. This is done to explore a wide range of contextual factors that impact family travel patterns, such as family histories of travel, as well as their various geographical and cultural locations.

Even when class relations might be playing an indirect role in terms of how families travel, it is important to note that most often family travel is undertaken for many other reasons than to prepare and groom their children for grown-up life. For example, family travel may be related to sustaining diasporic links (Coles & Timothy, 2004), seeing religious sites or going on pilgrimages (Norman, 2011), or generating or maintaining social networks within and across distances (Urry, 2003). Some of the families that appear in this book are frequent travellers – both for leisure, work, seeing family and friends, or maintaining ties with national relations and identities – while others have very limited travel experience. A closer focus on similarities and differences across social and national location might help elucidate more clearly how individual family travel practices navigate processes of globalisation, national social class structures, increased environmental awareness, as well as travel restrictions, such as those imposed by a pandemic (such as COVID-19) or particular national policies determining who may enter a country and who may not.

Family Travel: The Practices of Holiday-Making

In addition to the more macro- and meso-focussed studies on national and socio-demographic patterns of spatial movement of people across national borders for the purposes of leisure travel, a body of literature on family travel zooms in on the experiences and motivations driving such travel.

Family travel can be defined as 'a purposive time spent together as a family group (which may include extended family) doing activities different from normal routines that are fun but that may involve compromise and conflict at times' (Schanzel et al., 2012: 3). The most common benefits attributed to leisure family travel in the scholarship include openness to others; learning other languages and about other cultures; exploring the world; and dealing with uncertainty (Wu et al., 2019). Today, with the growing number of households in which both parents are employed, and the creation of 'spare time' becoming increasingly limited, family travel, for vacation or other purposes, has become a symbolic act seen to promote bonding and family togetherness (Durko & Petrick, 2013). Moreover, among parents it is widely believed that travelling with children can enrich their geographical, cultural, and linguistic knowledge about places visited. These forms of educative travels include visits to historical sites, seeking culturally enriching experiences or visits to unexplored places or locations connected to a community's diaspora (Li et al., 2020; Yemini & Maxwell, 2020). Yet, to date, limited scholarship has explored the extent to which family travel is largely 'educative' in its purpose or embedded in other motivations – such as family bonding time, connecting with extended family, or 'simply' rest, relaxation, and even hedonism. Extending this area of investigation is one of the key foci for this book (see Chapter 4).

In his seminal study on tourism as a sociological inquiry, Urry (1990) argued that experiences during travel are given a particular and unique interpretation, due to their occurrence in the specific context of leisure travel. He suggests that otherwise mundane experiences are interpreted as exotic (understood as unusual or different from one's everyday life) when occurring during the holiday-making process (Urry & Larsen, 2011). This process is constructed in a particular way, fuelled by the tourism industry and peoples' own desires. Urry (1992) later proposed that 'the tourist gaze endows the tourist experience with a striking, almost sacred importance' (p. 173). As early as 1990, Urry addressed, in detail, the importance of various forms of documentations, mainly photography, and the role they would come to play in modern forms of tourism, where appearance or absence of other people (in particular other tourists) played a role in sustaining a place as a worthwhile destination. It is important to note that family travel can (and often does) include activities that may not fall under Urry's depiction of leisure travel and tourism, such as visits to family and friends, religious sites and tours focussed on a specific subject matter like arts, history, and nature. These activities can differ to the extent to which the destination is considered exotic by the travellers, the motivations and purposes of the trips, and the ways in which they are shaped and impacted by the tourism industry. Thus, the various travel forms partially identified by previous research need to be further documented, and critically, we need to develop a broader and more flexible framework for analysing travel than that offered by the conceptual framework of tourist gaze. While Janta et al. (2015) offer five rationales for travelling, these kinds of insights do not offer deeper or firmer theoretical structures within which to understand these rationales or to explain differences between groups found. Therefore, our book seeks to offer such a framework for the future study of family travel. In the subsequent chapter we initially map out our starting theoretical frame for analysing family travel, and later on in the book (in Chapter 10), we review and extend this theoretical framework for future studies of family travel.

By studying the processual perceptions of families with regard to travelling, it becomes possible to understand how interactions with infrastructures of mobility differentially shape travel practices. For example, Radic (2019) divided the travel experience into three distinct phases: pre-travel decision and planning phase; the travel engagement phase; and the reinterpretation and evaluation phase. In each of these phases, families shape and reiterate their travel experiences. Air travel, in general, and each of the phases, in particular, are circumscribed by several constraints and opportunities. International travel involves legal constraints such as passports, visas, and permits of stay, where some people are able to travel more freely than others (Harpaz, 2019). Additionally, access to the mode of transportation, availability of funds to travel, skills to access information regarding travel, sufficient time for travel for all family members, and personal preferences on the desirability of travel, are all factors taken into account by Kaufmann, Bergman, and Joye (2004)'s concept of *motility*. This will be further introduced in the subsequent chapter, as it is a key concept drawn on in our analyses.

The Paradox of Mobility: Environmental Concerns and Cosmopolitan Capital

Although social sciences research has shown that the potentials for being mobile, has certainly increased significantly in recent decades, Kaufmann et al. (2018) are among those who have argued that 'mobility choices are much more limited than they seem' (p. 199). Lack of necessary economic resources; concerns about safety and access to adequate medical care or equitable legal systems; and stresses of everyday life, among others, significantly restrict travel possibilities across the world. Travel restrictions for particular passport holders, as imposed during the COVID-19 pandemic (Chinazzi et al., 2020), or due to advice of a country's foreign office, for instance, directly limit freedom of movement and the ability to aspire to travel to certain parts of the world. COVID-19 has certainly, at least for some, led to temporary or perhaps permanent changes in attitudes to, and aspirations for, travel. A current, and perhaps even more pressing challenge, which we will tackle later on in the book, is related to how families are engaging with concerns about global warming and environmental degradation as a result of the increase in mass travelling.

The past decades' enormous increases in international travel have given rise to concerns over the harmful effects of mass tourism and travel transportation on the climate (Morten et al., 2018; Spasojevic et al., 2018; van Birgelen et al., 2011). Due to the hazardous effects of airplanes' high rates of CO_2 emissions into the atmosphere, air travel is currently considered a significant contributor to global warming (Higham et al., 2019). It is estimated that aviation may account for 15–40% of global carbon dioxide emissions by 2050 (Dubois & Ceron, 2006). Since millions of people, and families in particular, choose to travel by air for leisure, the question of pro-environmental (emission-diminishing) behaviour has become an increasingly relevant topic of investigation in recent years.

Although the numbers are still small, a new trend in family travelling is starting to emerge where families exchange the usual mode of holidaying, which involves multiple short breaks accompanied by one longer holiday every year, with various kinds of pro-environmental behaviours to mitigate their environmental footprints when travelling (Hares et al., 2010; Higham et al., 2016). Among such pro-environmental behaviours we find everything from full abstinence from air travelling or only holidaying domestically; significant, or partial diminishing of travel by air; shifting to more environmentally friendly modes of transport; distance/location related calculations; and the use of various carbon dioxide offsetting schemes. The latter, in particular, has been criticised and dismissed as merely measures and mechanisms of alleviating guilt (Mair, 2011), and critical scholars have argued that an overwhelming academic focus on individual pro-environmental consumption behaviours takes focus away from the (unregulated) political and systemic conditions sustaining the still growing aviation industry (Higham et al., 2019; Shove, 2010).

The vast amount of scholarship on the environmental consequences of travelling, however, emphasises the general lack of significant behavioural changes to travel, despite awareness of, and concerns about, the climate (e.g. Higham et al., 2014; Kroesen, 2013). A study of UK holiday travellers, for instance, revealed that price, weather, family and friends, minimal travel time, and activities were among the most common considerations informing holiday-related travel, leaving little space for environmental concerns to impact travel decisions (Hares et al., 2010). Even among those who consider themselves environmentally conscious, little evidence supports the assumption that pro-environmental attitudes has led to more sustainable travel behaviours. For scholars in the social sciences working within the area of environmental behaviour, this paradox, also named the attitude-behaviour gap (Anable et al., 2006), has generated significant amounts of research. Alcock et al. (2017), for example, found that there was no statistical correlation between individuals' environmental attitudes, concern over climate change, and their propensity to take non–work-related flights, or the distances flown by those who do so. Some scholars have explained this discrepancy between intentions and practices by pointing to the common perception of holidaying as a right for all (Barr et al., 2010). Since the UN recognised the 'right to rest and leisure including […] periodic holidays with pay' as a human right in 1948, governments, especially in Europe, have included holidays in national measures of relative poverty (World Bank, 2020). The framing of holidaying as an entitled periodic 'rest', might explain the common distinction that many people make between holidays and everyday life where individuals tend to be more willing to engage with environmental concerns (recycling, cutting down on consumption, etc.) (Becken, 2007). A similar distinction was found in Lassen's (2010) qualitative study on work-related air travel, where environmental concerns were de facto excluded from the informants' narratives, even by those who were otherwise environmentally conscious, as they prioritised considerations of time, money, and comfort.

Although significant academic efforts have gone into investigating the relationship between pro-environmental attitudes and voluntary reduction of air travel (see van Birgelen et al., 2011; Büchs, 2017; Morten et al., 2018), limited research has empirically engaged with the question of how and why pro-environmental values might differ across social classes or groups, and in what way pro-environmental discourses on travel are negotiated among family members. The latter question is one we directly investigate further in Chapter 9 on flight shame. Empirical findings from this book imply that flying has become deeply embedded in most middle and upper socio-economic groups residing in the Global North. Flying has, in other words, become a salient component of modern forms of leisure time activities as well as being integral to identities. This makes it a sacred practice, resilient to change – even the kind needed to mitigate the hazardous effects on the climate it is causing.

Moreover, the increased political attention given to the negative effects of air travel poses an interesting dilemma for middle- and upper-class families, especially those who regard themselves, or are seeking to distinguish themselves, as cosmopolitan (Maxwell & Yemini, 2019). Accumulating social and cultural capitals as cosmopolitans through spatial mobility and a global outlook, cosmopolitan families across the world, to greater or lesser extents, have to grapple with the paradox that cosmopolitan identities are often associated with and characterised by pro-environmental, global citizenship related dispositions (Adey et al., 2007). For most people, the development of a cosmopolitan orientation can only be achieved with frequent and long-distance air travel – seeing and experiencing distant places and learning to feel comfortable in them. But if, increasingly, a commitment to cosmopolitanism and global citizenship requires a deep commitment to environmental sustainability – how can this be married up with the need and desire to travel to be a true cosmopolitan? We examine this paradox in greater detail in Chapter 8. Furthermore, as Randles and Mander (2009) have argued, when frequent international travel by air became a possibility for the masses, it arguably also becomes less desirable for its assuredness in its ability to accumulate cultural capital and be drawn on during practices of distinction. To consider how this paradox is taken up and navigated by one social grouping, we examine the practices of global middle class families in Chapter 6.

***

Throughout the analytical and empirical chapters of this book, we will continue to engage with the strands of literature presented here, as we delve into the stories, experiences, opinions, and stances shared with us by our participants and reflected in our other sources of data. As we have established, the motivations, justifications, and patterns of travel can vary between different families and groups, and there is a need to grapple with the question of how the acquisition of cosmopolitan capital (be this a capital in its own right or a form of cultural capital – Maxwell and Aggleton, 2016), achieved through travel, is perceived to clash with environmental consciousness. Such variabilities lie at the heart of our analyses, which aim to comparatively explore how these issues are reflected in the travel patterns of families from different backgrounds and places, and the role played by values, culture, and status in shaping such patterns.

References

Adey, P., Budd, L., & Hubbard, P. (2007). Flying lessons: Exploring the social and cultural geographies of global air travel. *Progress in Human Geography*, *31*(6), 773–791.

Alcock, I., White, M. P., Taylor, T., Coldwell, D. F., Gribble, M. O., Evans, K. L., & Fleming, L. E. (2017). Green on the ground but not in the air: Pro-environmental attitudes are related to household behaviours but not discretionary air travel. *Global Environmental Change*, *42*, 136–147.

Anable, J., Lane, B., & Kelay, T. (2006). *Review of public attitudes to climate change and transport: Summary report*. Report commissioned by the UK Department of Transport.

Appadurai, A. (1996). *Modernity at large. Cultural dimensions of globalization*. London.

Aybek, C. M., Huinink, J., & Muttarak, R. (2015). Migration, spatial mobility, and living arrangements: An introduction. In C. M. Aybek, J. Huinink, & R. Muttarak (Eds.), *Spatial mobility, migration, and living arrangements* (pp. 1–22). Springer.

Balomenou, N., & Garrod, B. (2019). Photographs in tourism research: Prejudice, power, performance and participant-generated images. *Tourism Management*, *70*, 201–217.

Barr, S., Shaw, G., Coles, T., & Prillwitz, J. (2010). 'A holiday is a holiday': Practicing sustainability, home and away. *Journal of Transport Geography*, *18*(3), 474–481.

Bauman, Z. (1998). *Globalization. The human Consequences*. Polity Press.

Bauman, Z. (2000). *Liquid modernity*. Policy Press.

Becken, S. (2007). Tourists' perception of international air travel's impact on the global climate and potential climate change policies. *Journal of Sustainable Tourism*, *15*(4), 351–368.

Brown, F., & Hall, D. (2008). Tourism and development in the global south: The issues. *Third World Quarterly*, *29*(5), 839–849.

Büchs, M. (2017). The role of values for voluntary reductions of holiday air travel. *Journal of Sustainable Tourism*, *25*(2), 234–250.

Butterfield, D. W., Deal, K. R., & Kubursi, A. A. (1998). Measuring the returns to tourism advertising. *Journal of Travel Research*, *37*(1), 12–20.

Campo-Martínez, S., Garau-Vadell, J. B., & Martínez-Ruiz, M. P. (2010). Factors influencing repeat visits to a destination: The influence of group composition. *Tourism Management*, *31*(6), 862–870.

Capella, L. M., & Greco, A. J. (1987). Information sources of elderly for vacation decisions. *Annals of Tourism Research*, *14*(1), 148–151.

Carlson, S., Gerhards, J., & Hans, S. (2017). Educating children in times of globalisation: Class-specific child-rearing practices and the acquisition of transnational cultural capital. *Sociology*, *51*(4), 749–765.

Castells, M. (2010). *End of millennium*. Wiley-Blackwell.

Chinazzi, M., Davis, J. T., Ajelli, M., Gioannini, C., Litvinova, M., Merler, S., …, & Viboud, C. (2020). The effect of travel restrictions on the spread of the 2019 novel coronavirus (COVID-19) outbreak. *Science*, *368*(6489), 395–400.

Coles, T., & Timothy, D. J. (2004). *Tourism, diasporas and space*. Routledge.

de Haas, H. (2010). Migration and development: A theoretical perspective. *International Migration Review*, *44*(1), 227–264.

Dubois, G., & Ceron, J. P. (2006). Tourism/leisure greenhouse gas emissions forecasts for 2050: Factors for change in France. *Journal of Sustainable Tourism*, *14*(2), 172–191.

Durko, A. M., & Petrick, J. F. (2013). Family and relationship benefits of travel experiences: A literature review. *Journal of Travel Research*, *52*(6), 720–730.

Fong, Y. L., Firoz, D., & Sulaiman, W. I. W. (2017). The impact of tourism advertisement promotional videos on young adults. *Journal of Social Sciences and Humanities*, *12*(3), 1–16.

Frändberg, L. (2009). How normal is travelling abroad? Differences in transnational mobility between groups of young Swedes. *Environment and Planning A*, *41*(3), 649–667.

Fullerton, J. A., Kendrick, A., & Golan, G. J. (2013). Strategic uses of mediated public diplomacy: International reactions to U.S. tourism advertising. *American Behavioral Scientist*, *57*(9), 1332–1349.

Gamlen, A. (2014). The new migration-and-development pessimism. *Progress in Human Geography*, 38(4): 581–597.

Giddens, A. (1991). *The Consequences of modernity*. Polity Press.

Grindlay, A. L., Ochoa-Covarrubias, G., & Lizárraga, C. (2020). Urban mobility and quality of public spaces: The case of Granada, Spain. *Transactions on the Built Environment*, *200*, 37–48.

Harari, Y. N. (2014). *Sapiens: A brief history of humankind*. Vintage books.

Hares, A., Dickinson, J., & Wilkes, K. (2010). Climate change and the air travel decisions of UK tourists. *Journal of Transport Geography*, *18*(3), 466–473.

Harpaz, Y. (2019). Compensatory citizenship: Dual nationality as a strategy of global upward mobility. *Journal of Ethnic and Migration Studies*, *45*(6), 897–916.

Higham, J. E., Cohen, S. A., & Cavaliere, C. T. (2014). Climate change, discretionary air travel, and the "Flyers' Dilemma". *Journal of Travel Research*, *53*(4), 462–475.

Higham, J. E., Cohen, S. A., Cavaliere, C. T., Reis, A., & Finkler, W. (2016). Climate change, tourist air travel and radical emissions reduction. *Journal of Cleaner Production*, *111*, 336–347.

Higham, J., Ellis, E., & Maclaurin, J. (2019). Tourist aviation emissions: A problem of collective action. *Journal of Travel Research*, *58*(4), 535–548.

Horst, H. (2006). The blessings and burdens of communication: Cell phones in Jamaican transnational social fields. *Global Networks*, *6*(2), 143–159.

Janta, H., Cohen, S. A., & Williams, A. M. (2015). Rethinking visiting friends and relatives mobilities. *Population, Space and Place,* *21*(7), 585–598.

Jucan, M. S., & Jucan, C. N. (2013). Gender trends in tourism destinations. *Social and Behavioral Sciences*, *92*, 437–444.

Jutla, R. S. (2000). Visual image of the city: Tourists' versus residents' perception of Simla, a hill station in northern India. *Tourism Geographies*, *2*, 404–420.

Kaufmann, V., Bergman, M. M., & Joye, D. (2004). Motility: Mobility as capital. *International Journal of Urban and Regional Research*, *28*(4), 745–756.

Kaufmann, V., Dubois, Y., & Ravalet, E. (2018). Measuring and typifying mobility using motility. *Applied Mobilities*, *3*(2), 198–213.

Kim, S. S., & Prideaux, B. (2005). Marketing implications arising from a comparative study of international pleasure tourist motivations and other travel-related characteristics of visitors to Korea. *Tourism Management*, *26*(3), 347–357.

Kroesen, M. (2013). Exploring people's viewpoints on air travel and climate change: Understanding inconsistencies. *Journal of Sustainable Tourism*, *21*(2), 271–290.

Larsen, J. E. (2004). De marginale rums politik. In H. S. Andersen & H. T. Andersen (Eds.), *Den mangfoldige* (pp. 213–230). Statens Byggeforskingsinstitut.

Lassen, C. (2010). Environmentalist in business class: An analysis of air travel and environmental attitude. *Transport Reviews*, *30*(6), 733–751.

Lucena, R., Jarvis, N., & Weeden, C. (2015). A review of gay and lesbian parented families' travel motivations and destination choices: Gaps in research and future directions. *Annals of Leisure Research*, *18*(2), 272–289.

Lucht, H. (2012). *Darkness before daybreak: African migrants living on the margins in Southern Italy today*. University of California Press.

Li, M., Xu, W., & Chen, Y. (2020). Young children's vacation experience: Through the eyes of parents. *Tourism Management Perspectives*, *33*, 100586.

Mair, J. (2011). Exploring air travellers' voluntary carbon-offsetting behaviour. *Journal of Sustainable Tourism, 19*(2), 215–230.

Markwell, K. W. (1997). Dimensions of photography in a nature-based tour. *Annals of Tourism Research, 24*, 131–155.

Maxwell, C., & Aggleton, P. (2016). Creating cosmopolitan subjects: The role of families and private schools in England. *Sociology, 50*(4), 780–795.

Maxwell, C., & Yemini, M. (2019). Modalities of cosmopolitanism and mobility: Parental education strategies of global, immigrant and local middle-class Israelis. *Discourse: Studies in the Cultural Politics of Education, 40*(5), 616–632.

McKay, D. (2007). "Sending dollars shows feeling" - Emotions and economies in Filipino migration. *Mobilities, 2*(2), 175–194. doi:10.1080/17450100701381532

Milman, A. (2012). Postcards as representation of a destination image: The case of Berlin. *Journal of Vacation Marketing, 18*, 157–170.

Morgan, N., Pritchard, A., & Pride, R. (2007). *Destination branding: Creating the unique destination proposition* (2nd ed.). Elsevier Butterworth-Heinemann

Morten, A., Gatersleben, B., & Jessop, D. C. (2018). Staying grounded? Applying the theory of planned behaviour to explore motivations to reduce air travel. *Transportation Research Part F: Traffic Psychology and Behaviour, 55*, 297–305.

Norman, A. (2011). *Spiritual tourism: Travel and religious practice in western society.* Continuum.

NurMuhammad, R., Horst, H. A., Papoutsaki, E., & Dodson, G. (2016). Uyghur transnational identity on Facebook: On the development of a young diaspora. *Identities, 23*(4), 485–499.

Poria, Y., Atzaba-Poria, N., & Barrett, M. (2005). Research note: The relationship between children's geographical knowledge and travel experience: An exploratory study. *Tourism Geographies, 7*(4), 389–397.

Poria, Y., & Timothy, D. J. (2014). Where are the children in tourism research? *Annals of Tourism Research, 47*, 93–95.

Poria, Y., Atzaba-Poria, N. & Barrett, M. (2005): The relationship between children's geographical knowledge and travel experience: An exploratory study. *Tourism Geographies, 7*(4), 387–397.

Prayag, G., & Hosany, S. (2014). When Middle East meets West: Understanding the motives and perceptions of young tourists from United Arab Emirates. *Tourism Management, 40*, 35–45.

Prayag, G., & Ryan, C. (2011). The relationship between the 'push' and 'pull' factors of a tourist destination: The role of nationality – an analytical qualitative research approach. *Current Issues in Tourism, 14*(2), 121–143.

Radic, A. (2019). Towards an understanding of a child's cruise experience. *Current Issues in Tourism, 22*(2), 237–252.

Randles, S., & Mander, S. (2009). Aviation, consumption and the climate change debate: 'Are you going to tell me off for flying?' *Technology Analysis & Strategic Management, 21*(1), 93–113.

Salazar, N. B. (2018). *Momentous mobilities. Anthropological musings on the meanings of travel.* Berghahn Books.

Schanzel, H., Schänzel, H., Yeoman, I., & Backer, E. (Eds.). (2012). *Family tourism: Multidisciplinary perspectives* (Vol. 56). Channel View Publications.

Shove, E. (2010). Beyond the ABC: Climate change policy and theories of social change, *Environment and Planning A, 42*, 1273–1285.

Simpson, P. (2011). 'So, as you can see…': Some reflections on the utility of video methodologies in the study of embodied practices. *Area, 43*, 343–352.

Singhal, A., & Rattine-Flaherty, E. (2006). Pencils and photos as tools of communicative research and praxis: Analyzing Minga Perú's quest for social justice in the Amazon. *International Communication Gazette, 68*(4), 313–330.

Spasojevic, B., Lohmann, G., & Scott, N. (2018). Air transport and tourism–a systematic literature review (2000–2014). *Current Issues in Tourism, 21*(9), 975–997.

Tjorring, L. (2016). We forgot half of the population! The significance of gender in Danish energy renovation projects. *Energy Research & Social Science, 22*, 115–124.

Tomić, S., Leković, K., Marić, D., & Paskaš, N. (2018). The role of children in family vacation decision-making process. *TEME, 2*, 661–677.

UNWTO (2018): A compilation of data on outbound tourism by country, including data on international tourism expenditure and outbound trips. https://www.unwto.org/country-profile-outbound-tourism (accessed on 3 November 2020).

Urry, J. (1990). *The tourist gaze: Travel and leisure in contemporary society.* Sage.

Urry, J. (1992). The tourist gaze "revisited". *American Behavioral Scientist, 36*(2), 172–186.

Urry, J. (2002). Mobility and proximity. *Sociology, 36*(2), 255–274.

Urry, J. (2003). Social networks, travel and talk. *British Journal of Sociology, 54*(2), 155–175.

Urry, J., & Larsen, J. (2011). *The tourist gaze 3.0.* Sage.

van Birgelen, M., Semeijn, J., & Behrens, P. (2011). Explaining pro-environment consumer behavior in air travel. *Journal of Air Transport Management, 17*(2), 125–128.

Van der Eecken, A., Spruyt, B., & Bradt, L. (2019). Giving young people a well-rounded education: A study of the educational goals parents attach to the leisure activities of their children. *Leisure Studies, 38*(2), 218–231.

Wallerstein, I. (1974–1980). *The modern world system. Capitalist agriculture and the origins of the European world-economy in the sixteenth century.* Academic Press.

Weenink, D. (2008). Cosmopolitanism as a form of capital: Parents preparing their children for a globalizing world. *Sociology, 42*(6), 1089–1106.

World Bank (2020). Synthesis of the human rights addressed in the planning and management tool for human rights based development projects, The World Bank & The Nordic Trust Bank. https://www.worldbank.org/content/dam/Worldbank/document/Human%20Rights%20Synthesis%20.pdf (accessed on 3 January 2021).

Wu, M. Y., Wall, G., Zu, Y., & Ying, T. (2019). Chinese children's family tourism experiences. *Tourism Management Perspectives, 29*, 166–175.

Yemini, M., & Maxwell, C. (2020). The purpose of travel in the cultivation practices of differently positioned parental groups in Israel. *British Journal of Sociology of Education, 41*(1), 18–31.

Ying, T., Norman, W., & Zhou, L. (2016). Is social class still working? Revisiting the social class division in tourist consumption. *Current Issues in Tourism, 19*(14), 1405–1424.

3 Thinking with Travel

Setting the Stage

In this book, we define travel in its broadest sense, as the movement of a person (in our case – families with children) from one geographical place to another. We focus mainly on relatively short-term (less than six months) and international travel (meaning outside of the current country of residence). By narrowing our focus to travel by families with children, we aim to examine more closely how processes of class-making maybe exercised through this practice and/or challenged across generations through travel-related choices made. Moreover, we look at how aspirations as well as concerns about travel are negotiated among family members, in those families, where the children are old enough to contest their parent's travel routines. In this book we show how families' travel practices are part of broader set of practices which can help to both illuminate social class position, but also intra-group particularities shaped by smaller differentiations between people. The similarities and differences we find in our data on differently located families (by social class, by geography, by culture) begin to offer insights into how mobility through travel shapes short-term experiences and longer-term future aspirations. In this chapter, we introduce the theoretical framework that has initially help to structure and facilitate our analysis. We combine several theoretical notions, including Kaufmann's concept of motility, Andreotti's typology of global mindedness, and Lareau's work on parenting strategies. But, as will be shown across the various chapters, new theoretical resources need to be drawn on to study newer types of questions and concerns in relation to family travel, and in other cases, our sensitising framework set out here is too restrictive to allow for a more exploratory analysis to emerge. We reflect on our theoretical engagements with family travel in the conclusion of this book.

Travel: as a Multifaceted Practice

In introducing his concept of the *tourist gaze*, Urry (1990, 1992) frames the act of travel as a meaning-making activity, that shapes otherwise mundane experiences and marks them and the spaces in which they unfold as exotic,

DOI: 10.4324/9781003056430-3

unique, and worthwhile. He discusses how other people shape our under-standing of certain places as exclusive or as 'must-see' places. Following this notion, Urry argues that our own and others' presence in specific locations, alters their meanings and the values we attribute to these places. Issues related to language, cultural cues and practices, aesthetic tastes, and activities avail-able are all important in shaping the ways families anticipate and experience travel. Furthermore, concerns about climate change and the need to practice religious observance are also critical when examining traveling experiences and the sense of ease or comfort experienced in the process. Sometimes a level of familiarness to the place or destination is sought after, but other times experiencing a lack of comfort is what is valued – 'feeling almost like home' vs 'feeling not at all like home'. Throughout the course of this book, we examine such experiences and their potential to generate meaning attributed not only to places and activities but also to the family's own identities and aspirations.

Travel by families is much more than that evocated by Urry's 'tourist gaze' (1990), or his further development of his seminar concept into a more critical examination of the possible dangers of tourism in Urry and Larsen (2011). Travels also include, among others, visits to see families scattered around the world, short-term mobility for education, and maintaining social ties with friends met while living in different global cities and locations. We consider these varied forms of travel, as they all play a role in helping to articulate particular priorities families have, capitals (or skills and experiences) they may be seeking to accumulate for their children and orientations they wish to cultivate. In these specific ways, travel/physical mobility, especially that which takes you outside the nation-state, could be understood as a perform-ative class practice.

Travel as a Performative Class Practice

In thinking about travel, Kaufmann et al. (2018) emphasise how 'moving in itself is not the goal. Rather, we move to engage in different activities and to maintain our social bonds' (pp. 198–199). Similarly, we can characterise leisure time family travel in the same way. Here, engagement with mobility acts as a signifier of certain capitals, networks, and discourses. It is a performative prac-tice exercised by parents with short-, medium-, and long-term outcomes in mind. In our book we examine how families, from a range of socio-economic positions and cultural backgrounds, travel abroad. We explore their reasons for travelling, where they go, how they travel, and what they hope to get out of this form of mobility. We seek to understand the extent to which family travel can be considered a parental cultivation strategy, and, subsequently, how social class location is constructed through the act of travel. In doing so, we assess whether travel practices can be considered easily differentiable across traditional social class lines. More than this, we look at how emerging trends related to travelling might challenge such patterns of travel.

Flamm and Kaufmann (2006) theorise mobility as an aptitude for movement and a form of capital; they argue mobility is not only about spatial movement but also about how such spatial movement is made possible through social position. Borrowing from Bourdieu's notion of class, understood as relational dispositions of agents and groups within a given social field (Bourdieu, 1987), Kaufmann et al. (2004) conceptualise physical mobility as something that not only affects an individual's or group's positioning within the social structure but may also itself affect social structures more broadly. To emphasise its association with a form of capital, Kaufmann et al. (2004) re-conceptualise mobility as 'motility' and utilise the concept of motility as a gateway for examining 'how entities access and appropriate the capacity for socio-spatial mobility' (p. 750). Kaufmann et al. (2018) define motility as a 'set of characteristics that enables people to be mobile, including physical capacities, social conditions of access to existing technological and transportation systems, and acquired skills' (p. 2). In particular, motility encompasses elements related to *access*, *competencies*, and *appropriation*. Access denotes the availability of transportation and communication infrastructures that enable travel as well as the necessary economic and social conditions. Competencies involve the physical, acquired, and organisational skills and abilities involved in making movement possible, including everything from health to permits to knowledge and planning abilities. Appropriation refers to individual's interpretation and willingness to 'act upon perceived or real access and skills' (p. 750), and relates to values, motives and habits of travelling (Kaufmann et al., 2004). Critically, Flamm and Kaufmann (2006) focus on both the potential and actual capacity for mobility – alerting us to the importance of aspirations in shoring up practices. Kaufmann et al. (2018) suggest that motility is a question of a combination between choice (or in other words – preferences) and resources (in other words – access and skills), and that the question of travel can never be reduced to the concept of choice itself.

Kaufmann and his colleagues' work on motility is fundamental to our study of family travel as it opens up questions about aspirations for, and ability to travel and considers how socio-economic-cultural location shapes these. Furthermore, drawing on a Bourdieusian frame of analysis allows us to consider travel as not only an individual family act but also as a set of practices that are generated and contested within communities of families. We use this to examine what is promoted as a 'worthwhile' or 'desired destination' within the families' social networks, and in turn, how individual desires and actions influence other families' choices. Thus, family travel practices are affected by social structures but also, in turn, affect social structures themselves – as the negotiation of values and meaning occur continually within the social space.

Different Orientations to Travel

Additionally, for our framework which seeks to examine differential travel practices, we require a way to theorise the different orientations families can have towards travel. This will allow us to further understand travel patterns as potential performative class practices, but also as practices that have a direct engagement with 'the global', understandings of the 'Other', and spatial negotiations of perceived 'distance' to other parts of the world. To help us integrate such an analysis into our data interpretation work, we draw on Andreotti et al.'s (2015) model of *global mindedness*. Andreotti and colleagues define global mindedness as a multidimensional construct that forms an individual's engagement with 'others and difference' in 'contexts characterised by plurality, complexity, uncertainty, contingency and inequality' (Andreotti et al., 2015, p. 254). Global mindedness is to be understood as a set of dispositions that denote an individual's 'embodied possibilities for action' (p. 255) providing the situational characteristics of the specific context. Andreotti and colleagues (2015) propose a spectrum of global mindedness against which we plot our families' different travel practices. Three distinct dispositions towards the world make up this spectrum: 'seeing the world' which is understood as a form of tourism; 'experiencing the world', denoting an empathetic disposition to travel and the experiences you have while being mobile; and 'being part of the world', also described as 'visiting'. 'Tourism' is associated with objectivism, where the world is seen in only one way. Meanwhile, 'empathy' is associated with relativism, where each individual is acknowledged as potentially holding different views and perspectives of the world, but people still understand others as different from him/herself. The third type of global mindedness is that of 'visiting' which,

> entails locating oneself in a different place, not with the ambition to think and feel like others in that place do, but to have one's own thoughts, feelings and experiences in a location that is different from one's own. A location where one is with and in the presence of others, exposed to the world, and open to being taught by unpredictable teachers and teachings.
>
> p. 255

It is further suggested by Andreotti and colleagues that theoretically tourism is connected with ethno-centrism, empathy with ethno-relativism and visiting with existentialism. In the analysis of our data, we highlight how these different orientations to global mindedness can be explicated via the narratives of travel undertaken by the families, highlighting the sometimes subtle differences between them, but also how the outcomes of these variations can be significant when observing the practices they shape.

Thus, Andreotti et al.'s work allows us to examine nuances in orientations expressed as part of family travel practices and the implications for this in terms of social relations across difference. Meanwhile, we combine this conceptualisation with Kaufmann's work on motility (skills, access, and intentions) so that we can make sense of who travels where, and why and how such mobility is facilitated. Additionally, it offers insights into the more strategic endeavour embedded within family travel practices in terms of signalling social class location and attempting to secure future social positions. In this way we can consider how international travel is a resource that might distinguish groups from one another and is used by parents in conjunction with other forms of capital accumulation to secure their children's social position, and make accessible different spaces – at local, national, and transnational levels (Maxwell & Yemini, 2019; Yemini & Maxwell, 2018; Yemini et al., 2019). We suggest that various dispositions for, and practices of, travel, serve as an important capital, accumulated by families to be later converted to other forms of capital (Erel & Ryan, 2019).

Travel as a Practice of Cultivation

In order to facilitate close analysis of each family's travel practices, we have lent on Lareau's (2003) concept of *concerted cultivation*. This work also draws on Bourdieu and his concepts of capitals and habitus, and his argument that social reproduction takes place first and foremost within families. Famously, Lareau argued that there is a classed cultural logic underpinning parental education strategies, where the middle and working classes appeared to practise their parenting in distinctly different ways. Lareau (2003) captured the distinct perceptions of the nature of childhood held by different social classes in the United States. She coined the term 'concerted cultivation' to describe the array of highly organised and monitored leisure and sport activities that were perceived by parents of the middle class as valuable for their children's future success. In contrast, members of the working classes in Lareau's research used 'the accomplishment of natural growth' as their main principle driving their child rearing. In this mode of parenting the children used their free time to explore the world by themselves based on the limits set by their parents and these children tended to establish deeper links with their family members, as they spent a considerable amount of time together. Lareau argued that these distinct forms of parenting led to prominent differences in the ways these children learned to act in the world. As stated by Harrington (2015, p. 473),

> Lareau concluded that the class differences in values and the 'cultural logic of childrearing' (2002, p. 748) lead to divergent paths for children: 'an emerging sense of entitlement' among children raised by middle-class parents and 'an emerging sense of constraint' (2002, p. 749) for children of working-class and poor parents.

In the same vein, mobility, or as Andreotti and Beck claim – the urge for cosmopolitanism and globalism, might present yet another middle-class endeavour, which celebrates detachment from a particular locality as desired and advanced. This, in turn, due to its facilitation by previously acquired economic and social resources, excludes the peripherally and marginally located working-classes from pursuing a similar 'project' (Farrugia, 2020).

Yet Vincent and Maxwell (2016) have argued that such a distinction between the differently resourced social groups in terms of intentions is perhaps less clear-cut. While access to resources may limit the extent, or the manner, to which parents can pursue their desires to 'cultivate' their child – the responsibilisation and intensification of parenting has been shown to affect all social groups and therefore their practices. In our current study we have therefore sought to examine closely the extent to which differently located social groups understand travel as a form of cultivation, and what resources they draw on to enact this. In doing so, we not only record who moves and who stays moored, but we also work to reveal the structures that make sense of these movements, or as Massey (1994, p. 149) puts it – articulating the 'power geometry of space–time compression'.

★★★

While travel has become normalised, even for those with relatively limited resources, we argue, based on our theoretical framework and the overview of the literature, that travel should function differently for families in various socio-economic-cultural locations. However, there has not yet been a comprehensive examination of the purpose of travel in families and how it is integrated into broader parenting and cultivation strategies. Furthermore, with the relatively recent experience of restrictions to travelling internationally, due to COVID-19 pandemic, and a growing imperative felt by many to consider the role of family travel in significantly contributing to climate change – how families understand and practice travel, and how they understand the tensions inherent in the decisions they make around travel, must be more closely investigated. We anticipate social class, as suggested by Kaufmann and colleagues and Lareau, will shape travel practices, but we are also cognisant that social class stratifications play out differently in various contexts and may not be as sharply serrated in many societies today. We therefore seek to examine the role of social class and context more closely, and consider how various resources and aspirations for mobility, and global mindedness, may nuance and make possible different engagements with geographical mobility in the form of family travel.

References

Andreotti, V., Biesta, G., & Ahenakew, C. (2015). Between the nation and the globe: Education for global mindedness in Finland. *Globalisation, Societies and Education*, *13*(2), 246–259.

Bourdieu, P. (1987). What makes a social class? On the theoretical and practical existence of groups. *Berkeley Journal of Sociology*, *32*, 1–17.

Erel, U., & Ryan, L. (2019). Migrant capitals: Proposing a multi-level spatio-temporal analytical framework. *Sociology*, *53*(2), 246–263.

Farrugia, D. (2020). Class, place and mobility beyond the global city: Stigmatisation and the cosmopolitanisation of the local. *Journal of Youth Studies*, *23*(2), 237–251.

Flamm, M., & Kaufmann, V. (2006). Operationalising the concept of motility: A qualitative study. *Mobilities*, *1*(2), 167–189.

Harrington, M. (2015). Practices and meaning of purposive family leisure among working-and middle-class families. *Leisure Studies*, *34*(4), 471–486.

Kaufmann, V., Bergman, M. M., & Joye, D. (2004). Motility: Mobility as capital. *International Journal of Urban and Regional Research*, *28*(4), 745–756.

Kaufmann, V., Dubois, Y., & Ravalet, E. (2018). Measuring and typifying mobility using motility. *Applied Mobilities*, *3*(2), 198–213.

Lareau, A. (2002). Invisible inequality: Social class and childrearing in black families and white families. *American Sociological Review*, *67*(5), 747–776.

Lareau, A. (2003). *Unequal childhoods*. Berkeley.

Massey, D. (1994). *Space, place and gender*. University of Minnesota Press.

Maxwell, C., & Yemini, M. (2019). Modalities of cosmopolitanism and mobility: Parental education strategies of global, immigrant and local middle-class Israelis. *Discourse*, *40*(5), 616–632.

Vincent, C., & Maxwell, C. (2016). Parenting priorities and pressures: Furthering understanding of 'concerted cultivation'. *Discourse: Studies in the Cultural Politics of Education*, *37*(2), 269–281.

Urry, J. (1990). *The tourist gaze: Travel and leisure in contemporary society*. Sage.

Urry, J. (1992). The tourist gaze "revisited": The concept of the gaze, the tourist, *American Behavioural Scientist*, *36*(2), 1–20.

Urry, J., & Larsen, J. (2011). *The tourist gaze 3.0*. Sage.

Yemini, M., & Maxwell, C. (2018). De-coupling or remaining closely coupled to 'home': Educational strategies around identity-making and advantage of Israeli global middle-class families in London. *British Journal of Sociology of Education*, *39*(7), 1030–1044.

Yemini, M., Maxwell, C., & Mizrachi, M. A. (2019). How does mobility shape parental strategies– a case of the Israeli global middle-class and their 'immobile' peers in Tel Aviv. *Globalisation, Societies and Education*, *17*(3), 324–338.

4 Parents Talking about Family Travel

This chapter was written in collaboration with Camilla Sofie Linander and Olivia Pauline Rud Mogensen.

Introduction

Why do families travel? Families travel in different ways and for different purposes. In this chapter we focus on this, as a way of opening up and deepening our understanding of the practice of family travel. We explore how parents come to decide on where to travel and what activities they will engage in while travelling, as well as looking more closely at what parents hope to gain from these practices.

We examine how parents 'do family' through travel and how parents understand travel as directly and indirectly linked to creating futures of their children (from potential educational trajectories they could pursue, to their anticipated mobility in later life for leisure and for work). We show how travel experiences are shaped by families' context and own stories, and we consider how the COVID-19 pandemic impacted family travel.

Theoretical Note

In this chapter we delve deeper into the role travel plays for 'seeing the world', developing an awareness of Others, an expectation of geographic mobility, and how parents draw on travel as a concerted strategy to accumulate capital for their children. To facilitate this endeavour, we draw on Lareau's (2003) concept of 'concerted cultivation', given that all the families discussed in this chapter can be said to be from the middle classes. Yet we anticipate that family travel is not always so instrumental, given its associations with leisure, hedonism, and an event which all members of the family would ideally like to benefit from. Thus, in this chapter we focus on understanding other ways in which parents engage with the role of international travel for their family lives, and take careful note of differences found across the samples, to allow us to nuance and open up the family practice of travel further.

Given our focus on 'family', we have also examined how Bourdieu's theorisations of habitus and its extension, argued to be theoretically coherent

DOI: 10.4324/9781003056430-4

by some scholars, into the concept of 'family habitus' might illuminate our findings (Forbes & Maxwell, 2019). For Bourdieu, the habitus is the product of history; history produces individuals and collective practices. The 'system of dispositions' constitutes 'a past which survives in the present and tends to perpetuate itself into the future …' (1977, p. 82). We seek to understand how family offers particular ways of knowing and being through travel, and about travel, and draw on the concept of family habitus to do so.

Methodological Note

The following analysis is based on two sets of data: one collected through in-depth research interviews with Danish parents of children under the age of 25 and the other through an open-ended survey with parents living in different countries of the Global North. In total the findings for this chapter have emerged from the analysis of the accounts of 55 parents (nine inter-viewees and 46 survey respondents). While the in-depth interviews took a less structured and more explorative approach to the question of travel and its multiple purposes and meanings, the survey questions focussed specifically on questions related to frequency and destination of family travel, as well as their relation to broader educational goals. Furthermore, as the data collec-tion occurred during 2020, the possible effect of the COVID-19 pandemic on families' travel trajectories was examined as well.

Findings

Both the interviews with Danish parents and the open-ended survey re-sponses from the middle-class parents residing in countries across the Global North revealed a high familiarity with the practice of family travel. For these parents, travelling took many forms and served multiple needs related to work, leisure, and maintenance of relationships with families and friends in previous and current 'home' locations. For most of the families, travelling had become an integral part of a habitual routine, providing a basis for sus-taining while continually expanding the physical as well as social arena in which their family lives unfolded. In the following we explore the norms underpinning practices around family travel for these two groups as well as the kinds of aspirations that families attach to travel.

Travelling as a Child-Centred and Habitual Family Practice

Regardless of the kind of travel, the tendency to emphasise the importance of children's experiences while travelling was evident across the vast majority of the families in this study. When asked who stood to benefit the most from their family travels, almost all the parents participating in the open-ended-survey pointed to their children. Some parents described how destinations and activities embarked on during their travels were specifically chosen so as to give their children a chance to explore, be stimulated, or advance their existing skills, interests, and orientations outwards to the wider world. Both groups often emphasised activity-packed travels, like skiing trips, surfing vacations, sailing adventures, camping trips, and other holidays involving a variety of physical activities like hiking, kayaking, climbing, and horseback riding. For these kinds of travels, parents tended to highlight the physical and health-related benefits for their children and stressed the desire to develop their children's *'adventurous sides'*, as one mother described it. This focus – on their children's development – was also true for travels more focussed on traditional forms of knowledge and cultural learning outcomes such as travels to historical sites or culturally unique destinations. Many parents considered language skills an important learning attribute of travelling. Moreover, some expressed a wish to inculcate in their children a greater awareness of the geography of the world – so that they could consider where and why they might like to travel to a certain destination, alongside the development of skills to be able to plan a holiday, including the many considerations they would need to be aware of, such as transportation possibilities, the need to manage on a budget, and learning to adapt and appreciate different kinds of accommodations available in different locations visited. The exposure to, and accumulation of, experiences and skills through travel to other countries, interaction with other cultures, and the realisation of the vastness of the world beyond their national borders which could be explored were understood to have immediate benefits as well as longer-term implications for the development of their children.

Similar reflections on the potential beneficial outcomes of travel for their children were found in the accounts of Danish parents. To Lise, for example, exposing her two children to historical sites and cultural knowledge was an important part of going travelling. Her family holidays often involved visiting cities that had historical significance or ones known for their museums. To Nina, showing her children the metropolitan cities of Europe was likewise considered 'part of their general education'. She too was keen to take her children to museums across Europe. Yet, unlike Lise, who prioritised more traditional cultural learning experiences, Nina and her husband emphasised the importance of passing on knowledge and skills relevant to living in nature, when going travelling. As a family they often spent their holidays exploring different activities in nature, such as making bonfires and swimming in forest lakes. Nina explained their decision to go to Crete with the children as follows: '… [W]e are interested in plants, so we simply wanted for the children to see some other climates – and some other plants'. For Henrik, a committed runner and father of two, sports and physical activity played a central role in his and his family's holiday choices. At least once a year, his family travelled to Playitas, a sports hotel on Fuerteventura, and he had participated in multiple marathons across the world (bringing his wife and children with him each time). In the cases of Lise and Henrik, the activities their families embarked on during holidays can be seen as a direct prolongment of the kind of work and hobbies they pursue in their everyday lives. Lise - being employed as a vice president at a Danish museum - her efforts to expose her children to historical knowledge during travel were suggestive of a tendency to use holidaying as a means to pass on her own interests and knowledge to her children. Henrik, meanwhile, worked as a professional salesperson of sportswear. Like Lise, he preferred to spend his holidays, combining his own interests with spending time with his family, thereby ensuring a common sphere of interests with his children. These three examples provide insights into how family travel can be used as a way to cultivate certain interests, skills, and values in children. Whether it is historical knowledge, skills related to living in nature, or building physically active lifestyles, holidaying is used by parents as a way of passing on practical experiences and providing their children with activity-based learning that extends beyond, but is often connected to, their everyday life. These forms of cultivation, led by what families felt were important and symbolised a constructed family identity, can be partially understood by drawing on Lareau's (2003) concept of middle-class parenting as a determined strategy of concerted cultivation.

Yet, while the accounts above portray an image of parents who were actively involved in deciding travel destinations and activities that would ensure enriching experiences for all family members, the interviews with Danish parents also revealed a tendency among some families to ascribe meaning to less structured forms of holidaying. As one mother, Lotte, described it,

> There are never any wild plans, when we go [on holiday]. So that's just the agenda. Someone might say "now I would like to do this". Then people join in and say "me too" [...] It's simply day to day or minute to minute. It's something that everyone contributes to.

For Lotte, travel activities should not be planned in advance, but ideally would arise as spontaneous initiatives, which made it possible for the entire family to contribute to making the holiday into something that was worthwhile for everyone. It demonstrated an inclusive approach to travel, in which the children were considered responsible contributors on par with the parents, and where the children were encouraged to take initiative and actively seek out the things that interested them. This mirrors previous findings on parenting strategies where placing the child at the centre of family life is prioritised (Stefansen & Aarseth, 2011). While the level of pre-planning and focussed activity differed across these accounts, at the heart of these narratives lay an emphasis on spending time together as a family – doing things together and creating joint memories. Such narratives seemed almost antithetical to a strategy of concerted cultivation, and yet, as described in more detail below, concerted energies were spent on creating emotionally laden memories, building a desire for future travel in their children, and engaging in the work of 'doing family'.

The emphasis on creating family memories infused with positive emotions was also integrated with some parents' specific reference to their intention to repeat or mirror some of their own holidaying experiences when they had been young for their children. These parents tended to want to bring their children to places or engage in activities which they themselves had had as children. Nina, for instance, had primarily travelled within Scandinavia during her childhood, and had spent many of her holidays in her parents' holiday cottage in another part of Denmark. Her fond memories of summers spent there and the strong sense of connection she felt to their holiday cottage and the place in Denmark where it was located were an experience she endeavoured to re-create for her own children. Thus, they holidayed there several times a year.

Seeking to re-create particular emotions associated with travel was also something Anna remembered from her childhood and sought to reproduce for her own children:

> I have been brought up with this routine where you pack your car and then you blaze off [...] It is hard to explain, but it is just a little more real holiday-like. I don't know if it is something you bring with you from childhood. A kind of sensation in the body. It is what I have been used to doing, you could say.

Anna's account illustrates how travelling is an emotional practice, subject to bodily sensations that go beyond reflexive intent and conscious aspiration, led by a form of nostalgia aimed at creating a sense of embodied belonging to a place (Yemini et al., 2020). Her difficulties pinpointing exactly what it is that makes travelling feel 'holiday-like', shows how family travel involves internalised habits and experiences following their own practical logics. Even as a grown-up, travelling with her husband and son, Anna continued to seek out travel experiences that made holidaying seem 'more real'. In doing so, she drew on her past emotional experiences from when she was a girl. This emphasis on the habitual (repeating particular holidays) as a core part of being able to create lasting memories through travel, as well as a desire to reproduce certain associations, memories, and emotions across generations through travel, highlights its deep embedding within family life as a practice that is not only a form of instrumental cultivation but also a strongly emotion-focussed endeavour related to notion of family connectedness.

Many of the accounts of travel from both groups of parents involved having to make compromises. This was often the result of a disagreement in travel desires between the two parents, having to make collective choices around where to travel and what to experience while travelling with their children. In the following account we see how efforts to integrate the wants and likes of both parents can result in multifaceted forms of family travel, or what Nina herself called 'packages':

> I think, they [her children] should get around and see Berlin and Paris and London and Rome and those kinds of things, right? But my husband is not […] into city vacations, so it sort of has to be packed together. Then we make some packages. Then we must mix it together. Then we must canoe in the Ardennes and then see Brussels. You know, it is possible to mix things up, so that it becomes a good package.

Nina's account demonstrates how holiday plans are negotiated, whenever families are forced to compromise their individual habits in order to ensure the incorporation of opinions and preferences of both parents. This allows for the forging of what Vincent and Maxwell (2016) have called a *family habitus*, where values, priorities, experiences, and inclinations of multiple individuals are merged into collective practices around travel. Thus, while the negotiations between conflicting travel practices are subject to reflexive and conscious discussions, the compromises around how to travel eventually come into play and become unconscious logics, according to which future family travels are planned. Eventually the conflicting elements are transformed into naturalised and seemingly consistent 'packages' of travel practices. In this way, family habitus plays a central role in determining the right and wrong ways of travelling. Various or singular activities may be pursued, family time on holiday may be very structured or completely unstructured, particular destinations and desired connections to place and emotions may be sought out, but over time – the family develop a particular approach to travel (often not only very similar to their childhood experiences but necessarily re-negotiated with their co-parent and with parents' current interests and passions), which, we argue, is integral to, and illustrative of, a family's habitus.

Even for those parents who stressed the novelty and exhilaration of experiences as their main purpose for travelling, there seemed to be a visible pattern to the way most parents travelled. Although travelling was often framed as something that allowed for new experiences and meetings with 'the unknown', in practice it was often rather routinised and predictable. Several parents made a habit of returning to the same destinations they had been to as children, hoping to expose their children to the same feelings of comfort and familiarity that they themselves attached to these places (as Nina did). Gitte explained how she intended to encourage her children to go backpacking once they were old enough, as this was a form of travel that had allowed her to have numerous valuable experiences in her own youth. Nevertheless, in all responses there was a desire to pass on a love for travelling and to emphasise its positive outcomes (excitement, familiarity, and learning new skills) to their children. Parents believed they should and would pass on to their children an understanding of what it means to travel and a commitment to its importance in their lives. As Lotte said, travelling with the family was a kind of 'culture' they created. This culture shaped the habitual practices of a particular family's travel. It served the function of stimulating certain interests, skills, and preferences in their children, that not only provided them with capabilities and preferences that matched those of their parents but also allowed for the family to spend purposive time together. This did not necessarily mean that travel routines stayed exactly the same, however. As children stood at the centre of their plans, as they grew older, holiday destinations and activities might change a little. Yet the importance of 'getting away' together as a family and pursuing agreed-on family interests remained at the heart of their travel planning and the evaluation mechanisms used to determine whether a holiday had been successful or not.

The Importance of 'Going Away' Together

A persistent narrative, characterising the interviews with Danish parents, revolved around notions of 'the right way to travel'. Often uttered without any effort made to explain its rationale, such implicit expectations and standards around how to travel the 'right way' helped to make visible the practical logic underpinning family travel. Although the informants had travelled to multiple destinations and engaged in a large variety of activities while on holiday, they all seemed to comply with the notion that for travel to be 'real', it had to occur beyond their national borders. Despite it being a common practice for these families to vacation in Denmark, they almost collectively understood travel as a practice that involved movement to foreign destinations. Some parents said that domestic vacations could not replace travelling abroad. Lotte, for example, said, 'I wouldn't want it [vacation in Denmark]; instead, I would like to do both. I think, it [going abroad] gives me something different'. Holidaying in Denmark was thus considered a complimentary form of vacation, that could not substitute the practice of travelling abroad. Similarly, Anna said,

> I wouldn't mind spending some weeks here [in Denmark] during the summer holidays [...] but I think I would always feel a kind of need to get away for a bit [...] So I don't really think I could settle for Denmark.

This 'need to get away for a bit' was often associated with 'going south', and this desire was explained with reference to how a good holiday entailed 'nice weather'. For the majority of interviewees, the level of success of a holiday could be measured against the quality of the weather while away. In the case of the Danish parents, spending summer holidays in Denmark was therefore assumed to jeopardise the chances of having a successful vacation, as the Danish summer weather could not be trusted to be warm and sunny.

However, there were also other, more implicit reasons for why travelling was associated with movement abroad. Travels to foreign destinations was seen to constitute a different scope of opportunities, allowing for experiences that could not be acquired in the home country. This was particularly evident in the survey responses. While skills and abilities like language, a sense of the world's geography, and sports can generally be said to represent known and valued learning goals of most school curricula across the world, many families expressed a desire to use family travel as a way to 'give them [their children] skills they can't acquire at home'. This included seeing 'non-industrial settings like their own' or exposing them to 'a wider set of lifestyles and professions'. Such experiences were associated with personal growth and enhanced social skills, and seen to foster 'broader socializing opportunities', that would 'flex their minds [to] adapt to diverse cultures'. As two of the parents completing the survey said,

> I love to see how my children embrace travel and enjoy the experiences. I think it makes them realise how easy it is to have different experiences, try new things, learn new skills, be tolerant of other cultures, and be curious. I think exposure to travel and the experience of living in a different country with different people opens their world view and opinions.
>
> Parent 1

> It broadens their mind by making them realize there are more than one way to do things and to think of things. Our kids are bilingual (English and Greek), and although they spent the vast amount of their lives in the UK, they consider Greece a home country too. They recognize and enjoy the Greek culture, language, customs, and history. At the end of every summer, they are psychologically ready to go to a Greek school there! Once they return to the UK, they keep on speaking to each other in Greek for the first few days.
>
> Parent 2

These accounts mirror the many other reports made by parents across the Global North who emphasised that the value of travelling was to inspire their children to become 'globally minded and culturally aware'. Among these parents, 'feeling at home in so many contexts' was considered an asset that not only invited feelings of 'tolerance' and 'sympathy' with people different from oneself but also involved the expectation that travelling would embed within their children a form of resilience with which they would engage with the world in their future lives, and that would bolster their confidence and ambitions.

Although the acquisition of skills and global mindsets played a role in almost every parent's account of family travel, 'going away' was also closely tied to the need for families to have some time for recreation and re-connection, which was considered difficult to obtain in their otherwise hectic everyday lives. The images they painted of holidays stood in stark contrast to their accounts of busy and overloaded daily lives. Thus, getting away represented itself as an opportunity to attain a feeling of being 'very present' with others in their family, that appeared so hard to do in their normal routines. Anna, one of the Danish parents in the research, for example, expressed that travel 'is a really good way to get away from home and relax, and not have to think about one's everyday chores. You can just be together in a different way'. Nina stressed that going away allowed her to escape the physical context of her home, which was a constant reminder of all the things that required her attention and which 'paralyses the joy of vacation'. Lotte explained how she saw travel as a way to get to know her children in greater depth, through conversations:

> [...] it's about being together. And having those conversations. Of course, I know them [her children], but I feel as if I am missing out on those conversations about their everyday life [...]. Or those deep conversations, where you listen more attentively. Where it is not like: "Did it go well?" and then I don't listen carefully [...] Where I don't have to take out the laundry or something.

In this case, travel served as a kind of liminal time and place where Lotte's family members could get to know each other more intimately without risking this process becoming disturbed by the normal responsibilities of everyday life. Aside from chores that took up parents' time at home, the holiday represented a contrast to the busy leisure life that many of the children led. This included sports and music activities, jobs, and playdates with friends. Despite the voluntariness of such activities, several parents expressed how these all worked to distance the family members from each other on a daily basis. Nina explained how she saw travel as an opportunity to put these activities on pause for the duration of a week or two, and ensure time for 'family' was found:

> [...] The great part about travelling is that we get to assemble the family. Because normally they [the children] attend scouts and band and 27 other things, right? And I do yoga [...] and the children have become very good at playing together [...] And that's one of the reasons why I love getting away, because I think it's a different frame. No one suddenly thinks that it's time for a playdate or something like that.

Nina's explanation demonstrates how the priorities that characterise the structure of their everyday life are the exact priorities that she attempts to deprioritise during their family travel. The simplicity and slow pace of going on holiday stands in stark contrast to the speed and multiplicity of activities that characterise the parents' portrayals of their everyday lives. This reflects Rosa's (2010) depiction of a neoliberal modernity, characterised by increasing acceleration and alienation. Despite unprecedented attempts to save time through ever-new technologies of communication and production, more and more people feel as if time is running faster, and that in order to stay afloat, they too have to pick up speed. Physically distancing the families from the context of their home, then, becomes a way of slowing down time, in ways that allowed for certain modes of being together as a family. This is further illustrated in the way some parents seemed to be able to let go of their work obligations, and the children appeared to reduce their use of social media, only when they crossed the Danish borders and entered a new country.

In these examples, holidays abroad were seen to bring the family together by virtue of minimising the physical 'escape opportunities' of family members, as Eva expressed it. Thus, going travelling becomes a way of limiting those structures of opportunity that usually serve to detach the family members from one other. Using Stefansen and Aarseth's (2011) notion of *enriching intimacy*, family travel can thus be seen as a way for parents to facilitate the emotional care labour that is expected within a family. Travelling abroad represents a platform for families to spend meaningful and intimate time together. It allows for the synchronisation of rhythms between family members whose lives are usually lived in 'parallel routines', as Gitte called it, detached from one another. Travel, then, becomes an important feature in the process of *doing* family, as the activities undertaken while away carry an educative purpose that extends beyond sheer entertainment, or instrumental capacity-building of the children. It allows for the development of emotional relationships between family members, at the same time as it introduces the children to a variety of activities, interests, and areas of knowledge, that are considerable foundational to a family's identity and therefore their habitus. Family travel is, in other words, an educative practice that is seen to enhance both the capitals or resources of the individual child, while at the same time teaching them about the value of relationships and how to become a meaningfully member of a family.

Whatever the purpose of going away, both groups of parents involved in our research emphasised the value of travel trajectories that extended beyond national borders. Travel to foreign destinations had come to represent a kind of liminal space, different from the everyday lives of families, serving as a perceived gateway into experiences that could not be acquired 'at home'. According to Urry (1990), one of the most fundamental features of tourism as a social practice is the notion of *departure*. With this, he refers to how the value obtained from tourist travel is derived from the 'limited braking with established routines and practices of everyday life', which allows 'one's senses to engage with a set of stimuli that contrast with the everyday and the mundane'. Thus, as this chapter has shown, 'to consider how social groups construct their tourist gaze is a good way of getting at just what is happening in the 'normal society'' (p. 2). Although the focus of this book is not limited to those forms of travel that can be considered tourism, we believe it is possible to apply Urry's idea to our analysis: that by focussing on what is different or liminal, we indirectly interrogate that which is normal as well. This section has shown how images of 'going away' are highly influenced by perceived limitations of what life at 'home' entails. This makes way for new connections and insights into the kinds of values and activities families prioritise, through the practice of travel.

Present and Future

In the survey responses, travel was sometimes described in rather open-ended terms as experiences that would eventually 'make them [the children] aware of their own place in the world', but often the children were also infused with more instrumental and tangible aspirations for the future such as getting accepted into educational institutions or acquiring skills relevant for certain jobs. Maybe the most overt was the example of a mother, who used holidaying to take her daughter on a 'college tour through the United States to visit potential colleges', to which her 13-year-old daughter might apply later in life. Another mother described how she expected her child's travel experiences to translate into a future educational setting: 'In the longer term, this [travelling as a family] forms in them a vocabulary of experiences, which is essential for making sense of the often abstract and decontextualized academic knowledge they will face in schools'. Similarly, other parents expressed hopes that going travelling now would inspire their children to attend international schools when older. More than its concrete usability, travelling was thus seen as an indirect way to strengthen children's ambitions to travel the world once they grew up, cultivate their aspirations, and inspire them to live, study, or work abroad.

While not all parents formulated their purposes in as strategic a manner as those mentioned above, it was a general trend among parents who had lived in multiple countries to openly articulate their educative aspirations on behalf of their children, when talking about family travels. This pointed to an overall expectation that children would 'learn to see the world differently', by 'exposing them to different kinds of trips and places'. Travelling was, in other words, considered a way of "broadening their horizons" and providing them with a set of 'open eyes' and an open attitude – a mindset of the traveller, which they would be unlikely to experience and acquire within the institutional educational settings of a national public or private school. In this way – and as was also found in the in-depth interviews with Danish parents – the function of travel can be not only instrumental but also to simply pass on a love for the practice of travel itself. The way parents talked about family travel revealed an overall assumption that travel in the present was closely tied to opportunities for their children in the future, relating both to tangible aims, like educational merits, and less instrumental aspirations, such as orientations to the world and a passion for travelling they would further pursue when they were older (and had their own children).

Given that the accounts for this study were collected during the time of a global pandemic, a topic recurring in both the interviews with Danish parents and survey responses was around how restrictions on travel were perceived to have had an effect the present and future of these families' travel plans. Among the interviews with Danes, the majority of the parents had spent their summer vacation (year 2020) in Denmark. Despite the overwhelming focus on the positive aspects of family travels to foreign destinations, presented in this chapter, many of the Danish parents considered COVID-19 to have brought about important insights in relation to travel. Among these was the ability to save money by not having to spend it on flight tickets, hotels, and visas. Others highlighted the positive impact that the pandemic had had on the climate, as the flight industry has been brought to a standstill. Tina, Anna, and Gitte considered this a good thing – that fellow Danes seemed to have had their eyes opened to the many vacation possibilities and holiday offers within Denmark.

A partial explanation for this perhaps surprising optimism about not being able to travel might be found in the fact that to most of these parents, the current limitations on movement were not considered long-term. Tina, for example, longed to be able to travel again soon: 'I'm looking forward to being able to travel again and getting out for a bit. To live rather than to simply survive'. Many other parents were convinced that they would be able to travel abroad again soon. However, there might also be another reason, which has to do less with what the individual parent deemed realistic, and more with the social expectations surrounding family travel. To Nina, the reason why staying at home over the summer holidays was considered tolerable, related to the fact that everyone else had done the same: '… it was fine by me, because our travel pattern suddenly fitted the mainstream. So, it was OK. I mean, the wildest thing anyone had done was the same as everyone else'. This revealed an important aspect of family travel: it showed how the evaluation of one's own travel patterns were negotiated in accordance with the perceived travel patterns of others. To Nina, knowing that no one else had had any wild travel experiences in the summer was what allowed her to enjoy her own holiday in Denmark. This illustrates that travel patterns are not isolated practices of individual families, but fall into a wider net of relationships about actual and perceived understandings of how others travel. This is an important feature of family travel, and something we will touch further on later in the book, when we look at how young people perceive travel in relation to their peers (Chapter 7) and when we discuss how young adults who experience flight shame negotiate their preferences in relation to their families (Chapter 9).

The disappointment about restrictions around travel were also emphasised by parents who responded to our survey, whose regular trips to see family that lived in other parts of the world had to be halted, and for whom regular leisure and cultural holidays structured their annual plans and sense of purpose. However, not a single respondent in the data drawn on for this chapter said that they intended to stop travelling abroad, once the restrictions had been lifted. This demonstrates the considerable challenge societies will have to fundamentally alter travel expectations, despite the significant environmental consequences international travel (by air) has. This finding also emphasises what fundamental and habitual travel is to so many families.

Conclusion

This chapter has explored the role and functions of short-term leisure travelling among Danish and other middle-class families residing in the Global North. Across a wide array of purposes and ways of travelling, three major conclusions can be drawn. First, that children were found to take a central role in the practices around family travel. Thus, while family travel is part of a parental strategy of child-rearing, individual parental interests and professional roles, as well as their own childhood experiences, were negotiated within the family and led to the development of a family habitus around travel. Second, 'going away' to foreign destinations was considered an integral part of being able to acquire the skills and experiences that travel afforded families, and these could not be assessed 'at home'. Just as importantly, however, which previous research has not examined closely or theorised, long-distance leisure travel was seen to offer families a liminal space that allowed them to spend meaningful time together. Thus, travel is not only a form of concerned cultivation (Lareau, 2003) but also a family practice that is focussed on enriching intimacy (Stefansen & Aarseth, 2011) and creating lasting memories that were to become habitualised and reproduced in the near future and across generations. Lastly, travel experiences were generally perceived to influence family members' possibilities for the future, thereby stressing the importance of travel, particularly for the children. Despite differences in travel patterns and the scope of aspirations related to travel, these tendencies largely persisted across the two groups of informants. This suggests that, although there are indeed many nuances to how and why families travel, the notion of family travel rests on certain assumptions, reflecting the social make-up underpinning modern societies and lifestyles across the Global North. While this chapter has provided a general introduction into how parents in different settings talk about travel, the subsequent chapters engage in more detail with how travel holds different functions and potentials for different families, depending on their social, economic, and cultural positions in society (Chapters 5 and 6), as well as how some families choose to reject conventional notions of travel and engage in alternative forms of family travel (Chapter 8).

References

Bourdieu, P. (1977). *Outline of a theory of practice*. Cambridge University Press.
Forbes, J., & Maxwell, C. (2019). Bourdieu plus: Understanding the creation of agentic, aspirational girl subjects in elite schools. In G. Stahl, D. Wallace, C. Burke, & S. Threadgold (Eds.), *International perspectives on theorizing aspirations: Applying Bourdieu's tools* (pp. 161–174). Bloomsbury.

Lareau, A. (2003). *Unequal childhoods*. Berkeley.

Rosa, H. (2010). *Alienation and acceleration: Towards a critical theory of late-modern temporality*. Aarhus: Aarhus University Press.

Stefansen, K., & Aarseth, H. (2011). Enriching intimacy: The role of the emotional in the 'resourcing' of middle-class children. *British Journal of Sociology of Education*, *32*(3), 389–405.

Urry, J. (1990). *The tourist gaze: Travel and leisure in contemporary society*. Sage.

Vincent, C., & Maxwell, C. (2016). Parenting priorities and pressures: Furthering understanding of 'concerted cultivation'. *Discourse: Studies in the Cultural Politics of Education*, *37*(2), 269–281.

Yemini, M., Maxwell, C., Koh, A., Tucker, K., Barrenechea, I., & Beech, J. (2020). Mobile nationalism: Parenting and articulations of belonging among globally mobile professionals. *Sociology*, *54*(6), 1212–1229.

5 Travel as Integral to Class-Making Practices

Introduction

According to Beck (2012), we are living 'in an age of cosmopolitization' (p. 7), where all relations – business, institutional, and familial – have become enmeshed and interwoven, to the extent that there is 'no outside' or 'other' anymore (p. 9), or at least this is true among certain echelons of the society (Brown, 2010). Beck (2012) claims that, in this era of 'second modernity' (p. 10), the traditional class structures that characterised the industrial nation-society have lost their salience as a mechanism for understanding social, economic, and cultural relations and stratification patterns today. Beck (2007) seeks to challenge this 'methodological nationalism' in thinking and to debunk the notion of social classes, which he considers a 'zombie category' (Beck & Willms 2004, pp. 51–52). He does this by introducing a 'cosmopolitan sociology', which addresses what he considers the new sub-political constellations of global society (Beck, 1996) while dismissing states' borders as natural units of analysis. Beck's proposition aligns with the image proposed by Bauman (1998) of 'globals' who move fluidly around the world, and for whom notions of 'home' and 'belonging' are not tied to nations. Yet most families do not move freely or continuously around the world, and find themselves locally moored (Yemini et al., 2019) in terms of their education choices and aspired-to futures (Maxwell & Yemini, 2019). Yet – as illustrated in the previous chapter – many families travel abroad for holiday, and seek out adventure, the accumulation of cosmopolitan experiences, and time together as a family. Thus, in this chapter, we wish to interrogate more specifically whether or not class might still have some relevance for understanding the resources, access, and skills differently located families display for holidaying abroad, and critically, what they see the purpose of such a practice to be. The findings in our previous chapter focussed mainly on middle-class respondents. Here we work with data generated through interviewing a broader range of families in terms of social location.

Considering that a significant proportion of the working-class families in this study shared immigrant backgrounds, a considerable amount of travel among these families was focussed on retaining links with their families and cultural heritage abroad. For these families, travel was seen as a means for

DOI: 10.4324/9781003056430-5

securing a good future for their children within the national context of their 'home' country, alongside playing a major role in their family-making practices. Although travelling among middle-class families was also considered a way of securing the 'right' futures for their children, these families demonstrated greater divergence in how this was to be achieved through being mobile for holiday purposes. Some saw travelling as a strategy that would guarantee their children a set of experiences and skills that would be transferable to an Israeli job market, while others saw travel as a means to underscore national and religious identities. We argue in this chapter that while class does not represent a holistic lens for understanding differences in travel aspirations, attitudes, and trajectories, it remains a key analytical tool when assessing how spatial mobility can determine prospects for social mobility.

Theoretical Note

To drive our analysis forward, we draw on Kaufmann et al.'s (2018) motility concept – as this acknowledges the role of social location in shaping geographic mobility – but we also integrate Andreotti et al.'s (2015) conceptualisation of global mindedness to examine how access to cosmopolitanism as a resource, skills and orientation might be differentially understood and valorised by our variously located respondents. In taking up the notion of global mindedness we argue that it is actually the links to the nation-state that are largely responsible for determining how travel experiences will be expressed and embodied by families. While Andreotti and colleagues borrow Hannah Arendt's (1968) metaphors of tourism, empathy, and visiting to describe the possible distinctions between the ways people might engage with strangeness, in our analysis we stress the role nation-states play in the imaginaries of parents. This role is prominent in how a cosmopolitan outlook might be nurtured by families. In Maxwell et al. (2020) we developed the concept of 'cosmopolitan nationalism' as an analytical lens through which to examine contemporary policies and practices that navigate global processes (such as school achievement ranking systems, human mobility, and people's future expectations for the kinds of skills thought necessary for an increasingly competitive and interconnected world), while considering the embedded nationalistic tendencies still present around citizenship and futures. Here we show how access, skills, and desire for mobility, as per Kaufmann's theorisation of mobility, interact with Andreotti and colleagues' understanding of global mindedness to actually articulate forms of cosmopolitan nationalism within families.

Working-Class Families: Sustaining Family Relationships While 'Seeing the World'

The group of working-class families in our Israeli sample was the most diverse in terms of ethnicity, immigration background, employment trajectories, and extent and destinations of family travel by air. All of the interviewees were either first- or second-generation immigrants from the former Soviet Union Republics or third-generation immigrants from Arab countries. Centring international travel around visiting and maintaining dispersed, transnational family links was a particularly pertinent focus for the working-class families' journeys in this part of our study. In our interviews we met a grandmother with her granddaughter who were flying to meet their family, as they did once a year (from Germany to Israel), a family of four flying to a family wedding (from Israel to London), families attending bar mitzvahs and funerals all over the world (in London and New York), and families flying on a special holiday to celebrate bar mitzvahs, honeymoons, and anniversaries (in London, Bangkok, and Budapest).

The working-class families we interviewed were especially keen to share with us their links to the Jewish diaspora, and the connections they had to various places. While it is extremely common for Jews from around the world to visit Israel for special family celebrations (Shoham, 2016), the parents in our study were actively reaching outside Israel to mark special religious or other kinds of family events. One family of eight, that we met at the airport, were all wearing white t-shirts with the picture of the bar mitzvah boy, on their way to celebrate the event in Budapest, the mother saying, 'We celebrating by travelling. Every special event is marked by a holiday in a special place'.

Air travel, in the case of the working-class families we interviewed, was also frequently mentioned as combining a holiday with cultural, religious, or familial duties. Some families upheld an annual routine of spending the summer in Russia with the grandparents or visiting Lyon (in southern France) to stay with the mother's sister and her family. Since air travel in these cases served to fulfil a complex combination of needs and wants, speaking in Andreotti's terms, it could be said to include all three global mindedness dispositions – of tourism, empathy, and visiting. That said, many of the responders constantly compared the places being visited with Israel, which was also a main topic of discussion between the parents and the children. The various advantages and disadvantages they perceived of life abroad were listed and critically reviewed, with considerable emphasis by these parents put on 'instilling the love for Israel' (a father of two on his way to meet part of his family living in Cyprus). Understanding the world through constantly comparing 'home' with other places led all parents to glorify Israel as a nation, a process reflected in the way the national education system seeks to ensure its students appreciate the possibilities made available through globalisation, but always embedding a strong patriotic commitment to the nation-state of Israel (Goren, 2020). Thus, as argued in Maxwell et al. (2020), just as many school systems engage in a practice of cosmopolitan nationalism, so too did our working-class Israeli respondents.

Alongside this practice of cultivation, the working-class families also used travel as a way of expanding their children's horizons. When explaining the reasons for their family travel, Ofra replied,

For me it is very important that the kids see the big world outside. One of the twins always asks me, "Mummy, how do you call that place that you suppose to go after the army?" and I always tell him – "University". This is how I am dedicated to make sure that they know what to do [in the future].

Here Ofra, without pausing for breath, links travel abroad with instilling an ambition to attend university in her children. Other working-class families in our study emphasised that international travel to visit family was also about 'seeing the big world', or as Hezi, the father in another working-class family phrased it, 'we'll open the world for them'. This 'seeing the world' was directly linked to fostering particular expectations for future, which in all cases was a desire to pursue academic educational trajectories within Israel.

Here, it appeared as if these parents were associating geographic mobility through travel as resulting in the accumulation of cultural capital, which would become embodied in the form of aspirations (Yemini & Maxwell, 2018). Kaufmann et al.'s (2018) concept of motility could be applied here to argue that mobility becomes a form of capital accumulation which activates the accumulation or access to other kinds of cultural capital – such as aspirations for higher education and future professional employment positions.

In our Israeli sample, another interesting opportunity for travel abroad was often made possible through the parents' workplaces – for example when the supermarket department they worked in hit its sales target, or when a photographer was sent to China to purchase new equipment. These kinds of organised travel opportunities were more popular among working-class families, but were not mentioned by those families belonging to the Israeli middle-class segment. For the latter group, travel was sometimes needed for their work, but seen as necessary as opposed to a 'reward' or 'treat'. Those working-class parents who mentioned travelling through work explained that such organised trips abroad meant they had to worry less about 'fitting in' or navigating unknown cultures (something they often expressed anxiety about) and therefore lowered feelings of travel-related uncertainty (e.g. being assured they would have access to Kosher food for religious families, or not being able to communicate in the local language).

To sum up, the working-class families we spoke to used travel to sustain cultural and familial relations across physical distance. This exposure to foreign settings and situations was linked not only to a concerted cultivation strategy by parents to extend their children's aspirations but also as a way for parents to embed their children's commitment to Israel as their 'home' and the place they would live in as adults. Ofra explained it thus:

I personally don't think that my kids should go abroad. This [Israel] is their country, they know the language, the culture and everything. The neighbours' grass is always greener as people say, I can't imagine us living somewhere else.

Ofra, despite being part of a widely dispersed family network, here seems unwilling to entertain the suggestion that her children might have more opportunities to secure a better future than she and her husband had in Israel. Israel was the space she wanted her children's futures to be played out in, but the opportunity to 'see the world' was a critical part of supporting her other concerted efforts (access to books, continually discussing appropriate future trajectories) to cultivate higher education ambitions and securing a professional job within Israel.

This is perhaps understandable given that most of the Jewish Israelis who immigrated from the former Soviet Union Republics and parts of the Middle East and Northern Africa had escaped terrible persecution, with Israel offering a safe haven for the first time in their lives. Thus, family travel met two key objectives: first, by exposing their children to the wider world, and by stimulating them through not only the thrill but also slight uncertainty these encounters engendered, parents believed their children's desire and confidence to strive for social mobility would be facilitated. Second, by combining an interaction with other places as a 'tourist' (using Andreotti et al.'s typologies of global mindedness) but also as a visitor (staying with family that lived in these other countries), the children, prompted by their parents' reflections, could gather cosmopolitan experiences, while simultaneously shoring up the importance and comfort Israel provided as their home nation. We suggest, in this way, family travel by our working-class Israeli families acts as a form of cosmopolitan nationalism, shaping and shoring up particular identities (a similar argument made by Wright et al. (2021) in their examination of Chinese young people attending non-traditional international schools). This is a very unusual, and undocumented parental strategy around aspirations and belonging, which could have significant effects, given that international travel is becoming more accessible to many. We also found a similar association made between travel and future aspirations for social mobility by young people from working-class backgrounds in Denmark, which we introduce in Chapter 7.

Middle-Class Families: National Identities versus Global Mindedness

For the middle-class participants in our Israeli study, family travel abroad was a more common practice than for those categorised as working-class, usually occurring two to three times a year, planned independently several months in advance, and not so closely tied to visiting extended family or celebrating particular family milestones. These middle-class Israeli families tended to travel for holidays to destinations more commonly associated with 'western Europe' (London, Paris, and Berlin), New York, and the islands of Thailand.

An Ease with Being Mobile: Towards an Emphatic Attitude to the World

All our families understood middle-class futures as having to engage with 'the global' to some extent – through work (employed by, collaborating, or trading with companies in other countries), through travel (for leisure or work), through meeting people from other parts of the world and using the common language of English to communicate, and through having to position oneself as someone seeking engagements with 'Others'. Such reference points could be associated with various articulations of cosmopolitan capital or cosmopolitanism as a form of cultural capital (Igarashi & Saito, 2014; Maxwell & Aggleton, 2016; Szerszynski & Urry, 2006; Weenink, 2008). In the following we show how the Andreotti's et al.'s (2015) model echoes the dispositions towards cosmopolitanism held by the different middle-class families in our sample.

One middle-class mother, Betty, explained her reasons for taking her children travelling as thus:

> We make sure to travel with them [the children] from a very young age, going on trips abroad, and of course within Israel as well. In Israel we go off on a regular basis, but also at least once a year we go abroad as well. The last trip was a trip to the Far East, to the third world, to Vietnam and Thailand, because I think it's very important, and even if they do not remember anything (and my little ones will not for sure) it teaches them that the world is big and there are many kinds of people, many ways to live a life. Some people look different, and make you feel different, that you are suddenly in a different place, I think that is important to the child.

Betty's description of their holiday plans and the articulated rationale for such trips abroad contain elements both of what Andreotti et al. (2015) describe as an *objectivist* and a *relativist* disposition towards the world. Betty suggested that the cultures in various 'third world' countries could be understood homogenously as a singular entity, and that it could be experienced and categorised as different to one's own. On the other hand, she emphasised that the learning experience she wished to make available to her children while travelling was to appreciate that there were a multitude of different ways to live one's life. Although such different lifestyles were still to be understood as essentially 'Other' from her children's own lives, this attitude expressed a certain level of what Andreotti et al. would typologise as 'empathy' towards the world, where lifestyles and places are compared without necessarily positioning them normatively within a clear value hierarchy. Learning to experience difference and to be comfortable with difference was positioned here as an important part of a middle-class child's education.

In order to learn how to navigate difference when visiting other parts of the world (whether for leisure in the current moment, or later on as a potentially necessary skill when conducting business with people in other parts of the world), English emerged as a core competency that middle-class parents in Israel wanted to gift their children. Ella, a middle-class mother, emphasised the need to cultivate English-language competencies through additional efforts at home but also as part of their holiday arrangements.

I am very strict on them with the English private tutoring, both of them regularly attend private lessons ... I see how important it is nowadays, and here in our neighbourhood there are many kids who have a very high level of English.

Ella mentioned that the family regularly chose to holiday in the United States, as well as sending her children to specific English-only summer camps in Israel and in the United States, and that the children practised their English when they met other English-speaking friends. Here, travel was seen complementary to the acquisition of certain skills needed for their futures, and integrated into the cultivation strategies already being put in place while at home.

Going Abroad to 'Polish' One's Identity

Tamir Levi and his wife Esti were travelling abroad with their three girls (ages 9, 7, and 2) for the first time since the birth of their youngest daughter when we spoke to them. Tamir's brother was working for one year with the Jewish community in a big European city and they planned to visit him and his family, but also to,

> wander around, as we planned this at the time of our annual family vacation (abroad) ... We will stay there for eight days, rent a car – will do: the Zoo, kids' museum and for two nights we will stay in a hotel in the forest, doing some attractions with the kids there.
>
> Tamir

As a family, the Levis perceived themselves as travelling less than other middle-class counterparts, whom they described as 'quite bourgeoisie' arguing that 'we believe in instilling values and not just running after the western beauty model or American lifestyle' (Tamir). For the Levi family, their travel practices served as a way of defining themselves in opposition to what they perceived to be the somewhat shallower travel trajectories of other middle-class families. 'We most probably do other things abroad ... not like the "typical" family. We don't have these check lists ... we would just walk in the park, spending time with the family', Esti elaborated. Such a process of distinction, which was only overtly expressed by the Levis from across our sample, was also mirrored in one narrative constructed by a young 14-year-old middle-class Danish women (Chapter 8) in our research. Although this did not appear to be very common, if mobility through family travel can be at least partially understood by Kaufmann's extension of Bourdieu and the introduction of motility – we anticipate that if we were to include a greater number of participants, more such processes of distinction would emerge in the construction of their narratives.

But it is also important to understand this process of distinction as not necessarily something that is done within similarly located groups, but may also be a process through which boundary-making is articulated. For the Levis, travel outside Israel was understood as enabling the cultivation of a strong understanding of themselves as Jews, and in particular being an Israeli Jew. While having some proficiency in a language besides Hebrew was seen as important – 'languages and cultures are the most important for me' and 'the girls are regularly having private tutoring in English', explained Tamir – his main concern as a parent was to educate his daughters that 'the right way to live is to be Jewish in Israel'. Other middle-class parents in this study identified a similar purpose of international travel. As another mother explained, 'we go out there, so the girls will understand who they are, develop their own meanings for experiences and then will be ready for all the challenges in this country'. Thus, it appeared as if for many Israeli Jewish middle-class families, travel became a means for embedding secure futures and identities in their home nation. Similar to the working-class families, middle-class families described in this chapter also practiced cosmopolitan nationalism, but their engagement with the 'Other' whom they met while abroad was more limited than those of the working-class families because they lacked the strong, often historical links to the societies visited. Even when Tamir's family visited his brother who was working in a European city on a two-year contract, Tamir's daughters would not have had a more authentic engagement with life in this city because their uncle was only a temporary resident. Therefore, the immigration backgrounds of our working-class families (an intersection that is almost synonymous with being working class in Israel) facilitated more of a 'visitor' engagement with the places visited abroad.

The Levi family had the resources and skills to travel and engage with the 'Other', but chose not to allow these encounters with the world to unsettle their affiliations to a particular articulation of being Jewish. The Levis primarily considered travelling to be about strengthening the family's outward identity as 'Israeli Jews'. Tamir elaborated further: 'Being abroad allows you to polish your identity. You can learn about the others to better know who you are. You can choose from different forms of Judaism that are out there'. With regard to future travels, Tamir offered, 'We will travel in the future as a family and we'll even consider visiting a developing country … again taking up the rucksack and seeing the various ways of life'. When asked what challenges they might face during more intrepid travels – 'You eat kosher only, right? How will you manage?' Tamir responded,

> it is challenging, but this is the way we are. We bring food with us, we know where to shop … we have our boundaries … I am not one of these people who think that the others' grass is greener … I know who I am.

This refrain by Tamir about knowing who he was the main driver shaping the Levis' travel plans.

Thus, Tamir and Esti's concerted cultivation strategies around travel focussed on ensuring that their daughters' strong sense of themselves as Jewish and Israeli became further embedded in their meeting with the outside world. This, according to their narrative, was done by exposing their children to transnational practices of travel, where 'Others' were temporarily encountered, but not actively engaged with (unless they too were Jewish). Here, the capital being accumulated via opportunities for international travel was another, subtly different type of cultural capital – that of identity and commitment to such an embodiment. Such a clearly articulated sense of self was seen as critical to navigating future key choices – around where to live, appropriate marriage partners, and so forth. Here, Kaufmann's motility (i.e. being mobile) enabled Tamir's daughters to better embody a sense of themselves as being from somewhere and belonging to a particular cultural and religious group; again we see this as a form of cosmopolitan nationalism being practised.

Conclusion

In this chapter we have explored the travel narratives and attitudes to 'the global' differentiated by socio-economic and socio-cultural location. Andreotti et al.'s (2015) typological framework has been useful in highlighting how different modes of global mindedness can reveal subtle but important differences in perceptions of the world, which inform why people travel. Nevertheless, the framework was not sensitive enough to record how different contexts might shape peoples' dispositions. We suggest that – first – the role of the nation-state needs to be further explored in family travel projects. Families, similar to institutions (Wright et al., 2021), were found to engage in practices of cosmopolitan nationalism (Yemini et al., 2020). Moreover, while all families employed travel for educational purposes through the promotion of cosmopolitan experiences and skills development, working-class families in our study were more able to develop a 'visiting' type of global mindedness, because they visited and stayed for reasonably long vacations with families permanently based in different parts of the world. An immigrant background is often perceived as a disadvantage in Israeli society, but in our research, we show that working-class families arguably have more opportunities to develop a deeper cosmopolitan engagement, due to family travel being focussed on visiting family in different parts of the world. Parents from both groups stated that developing a strong and lasting sense of belonging to Israel was not only performed locally but had to be extended and further embedded through travel abroad.

Not all societies are constructed in the same way as Israel, where the working classes are largely constituted by having an immigrant background from particular parts of the world. But in many countries similar associations could be found, such as in the United States, France, and Australia. It would be fascinating to examine whether, due to their diasporic links, these immigrant working-class communities could be argued to have a stronger cosmopolitan engagement, closer to Andreotti et al.'s 'visiting' disposition, than their majority ethnic, middle-class peers. Second, it would be an interesting extension to our findings to understand more clearly which immigrant communities and which nations appear to engage in a practice of cosmopolitan nationalism rather than a more nostalgic form of nationalism they practised in the countries they left behind (Yemini et al., 2020).

Acknowledgement

This chapter draws on data and arguments already partially developed in Yemini and Maxwell (2020). The purpose of travel in the cultivation practices of differently positioned parental groups in Israel. *British Journal of Sociology of Education*, *41*(1), 18–31.

References

Andreotti, V., Biesta, G., & Ahenakew, C. (2015). Between the nation and the globe: Education for global mindedness in Finland. *Globalisation, Societies and Education*, *13*(2), 246–259.

Arendt, H. (1968). *Between past and future: Eight exercises in political thought*. Penguin Books.

Bauman, Z. (1998). *Globalization: The human consequences*. Polity.

Beck, U. (1996). World risk society as cosmopolitan society? Ecological questions in a framework of manufactured uncertainties. *Theory, Culture & Society*, *13*(4), 1–32.

Beck, U. (2007). The cosmopolitan condition: Why methodological nationalism fails. *Theory, Culture & Society*, *24*(7–8), 286–290.

Beck, U. (2012). Redefining the sociological project: The cosmopolitan challenge. *Sociology*, *46*(1), 7–12.

Beck, U., & Willms, J. (2004). *Conversations with Ulrich Beck*. Polity Press.

Brown, G. W. (2010). The laws of hospitality, asylum seekers and cosmopolitan right: A Kantian response to Jacques Derrida. *European Journal of Political Theory*, *9*(3), 308–327.

Goren, H. (2020). Students in service of the state: Uncoupling student trips abroad and global competence. *International Journal of Educational Development*, *77*, 102226.

Igarashi, H., & Saito, H. (2014). Cosmopolitanism as cultural capital: Exploring the intersection of globalization, education and stratification. *Cultural Sociology*, *8*(3), 222–239.

Kaufmann, V., Dubois, Y., & Ravalet, E. (2018). Measuring and typifying mobility using motility. *Applied Mobilities*, *3*(2), 198–213.

Maxwell, C., & Aggleton, P. (2016). Creating cosmopolitan subjects – the role of families and private schools in England. *Sociology*, *50*(4), 780–795.

Maxwell, C., & Yemini. M. (2019). Modalities of cosmopolitanism and mobility: Parental education strategies of global, immigrant and local middle-class Israelis. *Discourse: Studies in the Cultural Politics of Education*, *40*(5), 616–632.

Maxwell, C., Yemini, M., Engel, L., & Lee, M. (2020). Cosmopolitan nationalism in the cases of South Korea, Israel and the US. *British Journal of Sociology of Education*, *41*(6), 845–858.

Shoham, H. (2016). 'He had a ceremony—I had a party': Bar Mitzvah ceremonies vs. Bat Mitzvah parties in Israeli culture. *Modern Judaism-A Journal of Jewish Ideas and Experience*, *36*(3), 335–356.

Szerszynski, B., & Urry, J. (2006). Visuality, mobility and the cosmopolitan: Inhabiting the world from afar. *The British Journal of Sociology, 57*(1), 113–131.

Weenink, D. (2008). Cosmopolitanism as a form of capital: Parents preparing their children for a globalizing world. *Sociology, 42*(6), 1089–1106.

Wright, E., Ma, Y., & Auld, E. (2021). Experiments in being global: The cosmopolitan nationalism of international schooling in China. *Globalisation, Societies and Education*, published online 2 February 2021.

Yemini, M., & Maxwell, C. (2018). De-coupling or remaining closely coupled to 'home': educational strategies around identity-making and advantage of Israeli global middle-class families in London. *British Journal of Sociology of Education, 39*(7), 1030–1044.

Yemini, M., & Maxwell, C. (2020). The purpose of travel in the cultivation practices of differently positioned parental groups in Israel. *British Journal of Sociology of Education, 41*(1), 18–31.

Yemini, M., Maxwell, C., Koh, A., Tucker, K., Barrenechea, I., & Beech, J. (2020). Mobile nationalism: Parenting and articulations of belonging among globally mobile professionals. *Sociology, 54*(6), 1212–1229.

Yemini, M., Maxwell. C., & Mizrahi, M. (2019). How does mobility shape parental strategies – a case of the Israeli global middle class and their 'immobile' peers in Tel Aviv. *Globalisation, Societies and Education, 17*(3), 324–338.

6 Global Middle Class Families

The Children as Active and Seasoned Globetrotters

Introduction

Initiated by the growth of multinational corporations, globally mobile professionals have emerged as a potentially new, social group. As employees they provide the expert knowledge and skills needed for the operation of such business entities by participating in global networks of production, consumption, and bureaucracy. According to Ball and Nikita (2014), these professionals are the servants to, but not owners of, the capital in these global networks; they can be described as what traditionally was considered middle class families. In other words, these families constitute a traditional middle and upper middle class in their countries, but they also practice extensive mobility, in a way that might eventually distinguish them from their moored communities, and thus potentially forming a social class of their own. These are highly skilled workers of diverse national origins who circulate the globe – mostly between key global cities such as New York, London, Berlin, Sydney, San Francisco, and Hong Kong – and serve as high-tech, financial, and legal specialists; middle managers; engineers; and other professionals (Ball, 2010; Beaverstock, 2005; Embong, 2000; Sassen, 1999, 2000; Sklair, 2002; Yeoh & Willis, 2005). The existing literature highlights three key features of this GMC group: (1) their frequent mobility, (2) a continuous negotiation around belonging, and (3) their particular strategies of cultivation in relation to their children, in an attempt to ensure their social mobility in life. In the following, we briefly present these bodies of research, in order to frame the discussion of these families' travel practices.

Theoretical Note

Frequent Mobilities

One of the most distinguishing features of these professionals is their pervasive mobility. Mobility means moving to live in one or more places outside one's 'home' country, having to travel regularly for work and liaise with professionals based across the world, while also necessarily maintaining social

DOI: 10.4324/9781003056430-6

and family networks virtually and through travel. As Favell (2008) found in his study of young professionals moving across Europe for work, mobility was articulated as 'liberating' them from some of the national social and cultural structures they perceived as oppressive back 'home'. Although Andreotti et al. (2015) argued that relationships to home might shift when these professionals must take into account the needs of children (and spouses), empirically we do not yet know enough about how the act of being mobile for work, the experiential aspect of learning to settle in a new space, and the expectation of continued mobility may alter global professional families' (parents' and children's) orientations to education, 'home'/their sense of belonging, and their anticipated futures. Furthermore, we need to understand how both geographical travel and mobility for work are drawn on as forms of capital. In this chapter we begin to consider this more closely.

A sense of Belonging

It has been argued that mobility shapes desires for the future and relations families have to the 'home' nation, the formation of social networks, and the seeking out of a sense of belonging for their children (Favell, 2008). It can be useful to consider the 'frames of reference' (Savage et al., 2005) parents draw on when determining the above – are they local (in which case, local to the geographic space in which they are currently residing, or to their home nation), or more global in scope? A global frame of reference invites the suggestion of a cosmopolitan or global citizenship orientation (Goren & Yemini, 2016, 2017), namely, of acceptance, interest, and comfort in engaging with the 'Other', but also potentially about being the 'Other' in a particular context. In such situations, elective forms of belonging (Savage et al., 2005) might be more likely to emerge, where the meaning of belonging is specifically developed by individual families to meet their own needs and desires. Drawing on the concept of boundary objects which are used to facilitate frequent boundary crossings, it was recently shown how promotion of language acquisition, and cultural or national rituals and traditions act as two central family practices that maintain strong connections to a form of national belonging for GMCs, despite being physically de-territorialised (Yemini et al., 2020). Missing to date, however, in our investigations into the GMC is whether and how cosmopolitanism is integrated with a sense of belonging, to shape the kinds of global mindedness dispositions engaged with.

Strategies for Success

Scholarly work highlighting the 'concerted cultivation' strategies that middle-class parents worldwide engage in (Irwin & Elley, 2011; Lareau, 2003; Nogueira, 2010; van Zanten, 2009; Vincent et al., 2012) is also relevant for our globally mobile middle-class families. Middle-class parents are argued to be ambitious in seeking to secure and extend their children's advantages, expose their sons and daughters to a broad range of extra-curricular opportunities, help them to identify their talents, and also develop them as a whole rounded person (Stefansen & Aarseth, 2011). As van Zanten (2016) argues, middle-class parents can draw on their own knowledge and personal experiences but also on professional experience of how to manoeuvre through systems of education to their children's advantage – conceptualised as a form of cultural capital. In the previous two chapters we illustrated how accruing cosmopolitan experiences was seen as necessary by middle-class parents for preparing their children for future opportunities. But how do future expectations of global mobility for one's own children, frequent re-location as part of a relatively regular occurrence, and regular returns to 'home' nations to visit family (i.e. these varied forms of mobility) get drawn on in terms of motility (a capital) in the concerted cultivation of practices of GMC parents?

In this chapter we specifically examine the co-construction of re-location mobility with holidaying mobility and its impact on (i) the accrual of capital in the form of motility and (ii) what kind of global mindedness dispositions are articulated through the cosmopolitan experience gained through mobility for the children of the GMC.

In what follows we will focus on three Israeli Jewish globally mobile families to show how travel is being utilised and perceived, further serving as a vivid illustration of the common class consciousness that is articulated. We integrate more general observations from our open-ended survey of global middle-class parents into these findings and our base of over 40 interviews with Israeli-origin GMC, in order to expand on or further qualify the analysis made based on the three family cases. We argue that motility (Kaufmann, 2014) is distinctively embedded in these families' practices and how it is anticipated it will shape their children's futures. Although, as Cairncross argued, as early as 1997, that distance is dead, the ways by which opportunities for international travel are taken up and cultivated offer another lens through which to analyse new stratifying mechanisms between social groups, despite access to travel becoming more equal. The routinised expectation of travel is deeply embedded in the parenting strategies of GMC groups (Beech et al., 2021) and makes them arguably distinctive.

Findings

In general, for our families that fall into the category of globally mobile professionals/GMC (educated to degree level and lived abroad for work purposes at least twice in the past 15 years), international travel for leisure, visiting friends and family, and short-term camps/study visits, all were deeply embedded within family routines. The children were seen as seasoned travellers, whose preferences were applauded by parents: 'He loves flying BA [British Airways]. Amazing how they know what exactly they want …' (as explained by one of the fathers in this sample). In the next sections we present a detailed account of each of the three families followed by a broader discussion.

The Naveh Family

The first family is the Naveh family. Our interview with the Naveh family helps to illustrate further the distinctive approaches GMC families took towards the act of travel and how it becomes an integral part of their cultivation practices for the present and future. For the so-called GMC families, living abroad and frequent re-location is a routinised practice (Ball & Nikita, 2014). The Naveh family was composed of the mother, Susanna (39), who held a postgraduate degree, had stayed at home with two children now aged 10 and 7 years for the last four years, but had recently started her own business importing South Korean health products to London. The father, Avi (42), also held a postgraduate degree in engineering, and was a mid-level manager at technology firm, based in London. They had an elder daughter and younger son. Susanna was born in Russia, raised in Israel and the United States, then returned to Israel, after which she studied in Paris. At the time of the interview, she had been living in London for the past four years due to her husband's re-location. Avi had been born and raised in Israel, had spent two years in the United States during his previous appointment, then returned to Israel for five years, before being transferred to London. All these professional re-locations were part of his employment with the same technology company.

Avi travelled frequently within Europe and occasionally (several times a year) to the United States. The family also travelled extensively for leisure and many times accompanied Avi during his work travels. Over the last couple of years, they have travelled to Italy, Canada, the United States, Seychelles Islands, and Germany. This image of a hyper-mobile family, that travels widely and routinely, was characteristic for all the GMC families that responded to the survey. Travels between a 'home' country – which can also be different to the parent's birth country – and either work- and/or holiday destinations with the entire family was a naturalised practice among these families. Going away five or six times a year for varying periods of time was not uncommon, whereas having two or three longer journeys planned for the year was almost considered mandatory. Travel in the Naveh family, too, was commonplace, and the children were encouraged to take a proactive role in planning the trips and deciding, comparing, and 'shopping' around for the best location and deal, as part of these preparations. 'Elena is the one who decides where to go, she is so independent. She knows how to compare on Tripadvisor and other sites and how to reserve places online'. Like the Naveh family, most of the other GMC parents tended to encourage their children to develop preferences and to take an active part in the process of choosing locations and activities prior to a journey. As one mother explained she and her husband would go on separate holidays with each of their two children so that the parents could get 'dedicated time with each child around his or her interests'. There was, in other words, an element of cultivating children to become independent co-producers of mobility within the travelling process.

Similar to working-class and middle-class families, the GMC parents tended to evaluate travel experiences based on the educative outcomes it had for their children. Although practical skills development – such as learning languages, acquiring historical and geographical knowledge, and attaining sports skills – was an important intended outcome for the GMC parents, most parents merely framed these skills as stepping stones towards developing personal preferences and a confidence that would enable their children to become seasoned globetrotters, capable of making sense of the world and interacting with all sorts of people regardless of nationality, class, and culture. In response to a question about choosing travel destinations, Susanna replied, 'we are not making a big deal about going away. It is not about having more stamps in the passport, and anyhow, the passports today are digitalised, it is about actually being part of the place'. This notion of 'being part of the place' was continued in the description Avi gave about his son's experiences abroad:

> We are there, but we send the kids to negotiate prices at the market by themselves, we would stay in places for a month so the kids get familiar with the place, he can play football with kids everywhere. This is part of the things that are so important, we have enough money to stay in fanciest places, but here (a village they visited in Vietnam), they feel at home. They can feel at home everywhere in the world.

Many GMC families in the survey expressed the hope that their children would develop some kind of 'psychological' and global 'mindset' that would allow them to 'flex their minds', 'adapt', and 'immerse' themselves into new cultural settings anywhere and interact with people everywhere. Similarly, other GMC families in our study wanted their children to '[be] free to choose'; 'feel at home in Goa and New York'; 'engage with people from different countries, cultures, classes and positions'; and 'take control of their lives'. These families also talked about the 're-location bug – once you did it, you want to do it again and again', suggesting that spatial mobility generated through travelling for leisure became a more ingrained aspect of their everyday and future lives. For these families, travel was understood as generating a sense of agency and adaptability that would enable their children to feel at home anywhere in the world. This was a specifically different orientation to travel as that expressed by other (working- and middle-class) families in our studies – who were hoping perhaps less ambitiously to 'open up the world' to their children, while securing their position 'back home'.

For GMC families, international travel was aligned to their broader school choice and informal education strategies – which focussed on being immersed in spaces full of 'diversity'. Susannah of the Naveh family, for instance, commented, 'the IB [International Baccalaureate] school … is amazing … You have huge diversity there, although people always move, they stay in touch, and have friends all over the world'. She continued later on in her interview to say,

[School] marks are important, but it really doesn't matter to me now. We are concerned with gaining experiences. When the kids will be older, they can choose from so many alternatives, and they will already have experienced flavours [of life] (from the various parts of the world).

Thus, for GMC families like the Navehs, travel (or spatial mobility in Kaufmann et al.'s (2004) terms) becomes integrated into their identities and future opportunities through frequent exposure. While this is arguably what differentiates them from their more locally moored middle-class peers, understanding how they narrate and integrate travel into their lives should offer further insights into whether or not mobility might be constitutive of these families emerging as a new kind of transnational social group. Certainly, for those working-class and moored middle-class families, spatial mobility was understood as advancing social position within their country, not just as a way of developing an orientation and a set of competencies that would allow their children to 'feel at home everywhere in the world'. Such an orientation is arguably closer to Andreotti et al.'s (2015) notion of 'visiting'.

The Peleg Family

The second family we introduce here is the Peleg family. The parents (two mothers; Hila and Ada) along with their three teenage children were interviewed in Germany in the midst of the COVID-19 crisis (spring 2020). The family had lived in Israel, then moved to Germany three years previously, and were planning to move again within Europe in two years' time. They travelled frequently, mainly to various locations in Europe, sometimes joined by friends and family. Travels had been a naturalised component of their family life. COVID-19 had, however, disrupted this routinised practice:

We think of travel as a fun activity. Every holiday from school we try to get away, to enjoy life as a family. The corona virus is not going to change that, but I do think we'll stay at home for the next few months, but hopefully we'll still be able to enjoy our carefully planned summer vacation in August.

Ada

The COVID-19 crisis and the decision by most governments around the world to implement restrictions on movement, offers a unique insight into the practices, expectations, and relationships that are at stake for those GMC families who consider mobility an integral aspect of family life. When asked how COVID-19 had affected her summer, Hila commented,

We had planned an extended family trip to Phuket in June (20 members) and the villa had been booked via Airbnb. We have to postpone the trip but perhaps my parents (who are in their 70s) may not be ready to travel next year if the situation doesn't return to normal. I won't be able to see my sister and niece who live in Melbourne. My parents like to travel nearby and to China and we do travel with them sometimes, perhaps we won't be able do it so often.

Ada, meanwhile, described her children as 'caged animals' under the corona restrictions. The pandemic has made clear the extent to which many globally mobile families take travelling for granted.

Likewise, for the majority of the GMC families we interviewed, visiting family and maintaining relationships with friends and relatives outside of their country of residence constituted a major component of their travel activities. As the Hila explained,

> We go back to Israel at least twice a year. For holidays, to see the family. For all big family celebrations, we go there. These links are very important for us. The children will probably live somewhere abroad in the future, perhaps study somewhere else, but they have very viable roots in Israel. A place to call home.

This concern (also mirrored in the other data we gathered with GMC families living elsewhere in the world and not of Israeli origin) – to be able to maintain close relationships to family across often large physical distances – was contingent on having the resources and freedom to travel. As another parent from the larger study commented,

> We might have our plan on going home for summer cancelled. That would be a disaster in many ways […] We'll be some of the first to travel when we can.

Drawing on Kaufmann's framework of motility, COVID-19 poses a threat to GMC families, because it dramatically affects the structures that enable and demand they travel across borders. Yet, as a mother in another family commented,

> It is hard to tell now, but we are not looking at the pandemic poetically. Once things are back to normal, we'll travel. Don't think we need to be panicking. Just because it is nice spending some time off together as a family.

This underlines the routinised and expected engagement with frequent travel GMC families have.

What further differentiates the GMC families from the working-class and middle-class families we interviewed was not that travel was used as a means to see families and friends (also critical for communities with large diasporic links), but that travelling and living in different parts of the world allowed their children to develop a critical perspective on their own and other cultures. As Ada explained,

> living in other places allows the children to really experience and be embedded in other cultures, being critical towards it, but also being critical towards their home culture. It develops you as a person, as a human being.

This emphasis differed from the testimonies of working- and middle-class families who tended to see travel as a means to develop one's own national identity, to have fun, and briefly experience other cultures. In other words, although travel is generally used as a way to sustain and strengthen relationships (within the immediate family, see the finding in Chapter 4; but also extended, transnationally located families, as for the GMCs and the immigrant working-class families in our studies), for the GMC – it is a way of life, a form of distinction, and a practice that promotes a 'visiting' global mindedness disposition, which aids children to become ever more adept at critiquing more nationally oriented senses of belonging. This stands in complete contrast to our Israeli working-class and middle-class families in the previous chapter.

The Heart Family

The third family we introduce here is the Heart family. The father, Ben, worked as a high-tech manager moving between major Silicon Valley–based global corporations, the mother, Karen, worked part-time in academia, and they had three sons aged 4, 6, and 12 years. The family had moved twice from Israel to London to take up new employment opportunities for Ben and were considering moving again in the future. But for now, they had been temporarily re-located 'back home' to Israel. The Hearts frequently travelled abroad, sometimes combined with travel for work. During their travels, the family usually visited museums, attended concerts and theatre productions, saw exhibitions, and went to shopping centres.

  Several of the GMC families we talked to regularly travelled back, not only to their 'home' countries or to where family and friends lived, but also liked to make a habit of going back to the places they had previously lived in. The was true for the Heart family who regularly travelled back to London, where the family has previously lived:

> After getting back to Israel, we would still travel to London a couple of times a year, just because we love it and miss it so much. We are not working on ticking off places and experiences, the children will have their time to further explore the world later on – for us the family experience is important. Not only London but in general we love to come back to places. It is more fun when you can get a real sense of the place, not just being a typical tourist.

Like the need and desire to maintain relational ties with friends and family 'at home', this tendency among GMC families to return to specific locations emphasises a preoccupation with creating meaningful relationships to specific places. As with the Naveh family, the Heart family described their approach to travelling as a way of getting 'a real sense of the place' (Ben) by contrasting it to what they considered 'typically touristic' modes of travel that often involved a 'bucket list' approach to deciding where to travel to next. The use of the word 'real' suggests a hierarchy between different ways of engaging with places, reminiscent of Andreotti et al.'s (2015) concept focussed on different ways of 'seeing the world'. Emphasising an authentic engagement with a place, enabled the Hearts to distinction between their travel practices to what they perceived to be a more superficial touristic consumption of place, by other families.

The Heart family had the necessary resources (money, the right kind of passport, language skills, and the employer paying for part of the family travel when it was work-related) that afforded them a significant freedom of movement. Thus, frequent travel was expected, and the focus was on ensuring their children would be able to gain specific orientations and individuality because of these naturalised and privileged family travel practices. Karen said, 'in the future we expect that the children will be able to express themselves through the experiences they gathered, not choosing to do something because everybody else is doing it'. Karen made this point with specific reference to being an environmentally aware and responsible consumer, because she felt that travel was central to teaching her children to become so.

Conclusion

Opportunities for, and the desire to, travel are very strong among the GMC families we studied. They appear to have an abundance of resources, skills, and access to practice short-term travel for holidays and other needs, which critically sits alongside their longer-term trajectories of mobility around the world for work. Although these families do have financial constraints, their orientation to mobility that is facilitated and strengthened by their work commitments means the purpose and possibilities for travel are significantly increased, when compared to the working-class and middle-class families in our studies. Moreover, GMC class-making strategies involve a strong reliance on the naturalisation of short-term family mobility, linking it to a broader set of attributes they seek to cultivate in their children, including patience, environmental consciousness, and intercultural competencies, among others. The GMC parents interviewed understood spatial mobility as an unquestionable naturalised right, and therefore travel as a relatively mundane activity, that had become deeply embedded in their families' routines.

The children in GMC families were encouraged from a very young age to develop preferences around travel, to actively participate in planning, designing, and negotiating their holidays. For these families, travel was seen as a valued and obvious cultural capital to accumulate. The GMC families sustained local links and relational connections with their 'home' countries, all the while maintaining a global outlook and set of cosmopolitan experiences that were to continue expanding their children's future horizons and 'visiting' orientations. While spatial mobility, acquired for most through holiday travel, was understood to cultivate motility as a resource for children that they could draw on in the present and future (i.e. mobility becomes a form of capital) for all families in our research, the articulation of this form of capital and its engagement with cosmopolitanism – as tourist, empathy, or visiting – differed according to families' different socio-economic and -cultural locations. In turn, this articulation of cosmopolitanism shapes the types of social mobility, identities, and sense of belonging being forged. A key question that remains, which COVID-19 has thrown up, is whether the distinctive positionings and lifestyles articulated by the GMC will be unravelled if unrestricted mobility never fully returns for this group.

References

Andreotti, V., Biesta, G., & Ahenakew, C. (2015). Between the nation and the globe: Education for global mindedness in Finland. *Globalisation, Societies and Education*, *13*(2), 246–259.

Ball, S. (2010). Is there a global middle class? The beginnings of a cosmopolitan sociology of education: A review. *Journal of Comparative Education*, *69*(1), 137–161.

Ball, S. J., & Nikita, D. P. (2014). The global middle class and school choice: A cosmopolitan sociology. *Zeitschrift für Erziehungswissenschaft*, *17*(3), 81–93.

Beaverstock, J. V. (2005). Transnational elites in the city: British highly-skilled inter-company transferees in New York city's financial district. *Journal of Ethnic and Migration Studies*, *31*(2), 245–268.

Beech, J., Koh, A., Maxwell, C., Yemini, M., Tucker, K., & Barrenechea, I. (2021). 'Cosmopolitan start-up' capital: Mobility and school choices of global middle class parents. *Cambridge Journal of Education*, 1–15. doi: 10.1080/0305764X.2020.1863913

Cairncross, F. (1997). *The death of distance: How the communications revolution will change our lives* (No. C20–21). Harvard Business School.

Embong, A. R. (2000). Globalization and transnational class relations: Some problems of conceptualization. *Third World Quarterly*, *21*(6), 989–1000.

Favell, A. (2008). *Eurostars and Eurocities: Free movement and mobility in an integrating Europe* (Vol. 56). John Wiley & Sons.

Goren, H., & Yemini, M. (2016). Global citizenship education in context: Teacher perceptions at an international school and a local Israeli school. *Compare: A Journal of Comparative and International Education*, *46*(5), 832–853.

Goren, H., & Yemini, M. (2017). Global citizenship education redefined–A systematic review of empirical studies on global citizenship education. *International Journal of Educational Research*, *82*, 170–183.

Irwin, S., & Elley, S. (2011). Concerted cultivation? Parenting values, education and class diversity. *Sociology*, *45*(3), 480–495.

Kaufmann, V. (2014). Mobility as a tool for sociology. *Sociologica 8*(1), 1–17.

Kaufmann, V., Bergman, M. M., & Joye, D. (2004). Motility: Mobility as capital. *International Journal of Urban and Regional Research, 28*(4), 745–756.

Lareau, A. (2003). *Unequal childhoods.* Berkeley.

Nogueira, M. A. (2010). A revisited theme—middle classes and the school. In M. W. Apple, S. J. Ball, & L. A. Gandin (Eds.), *The Routledge international handbook of the sociology of education*, 253–263, Oxon: Routledge.

Sassen, S. (1999). Whose city is it. *Sustainable cities in the 21st century, 16*, 145–163.

Sassen, S. (2000). Spatialities and temporalities of the global: Elements for a theorization. *Public Culture, 12*(1), 215–232.

Savage, M., Bagnall, G., & Longhurst, B. (2005). *Globalisation and belonging.* Sage.

Sklair, L. (2002). The transnational capitalist class and global politics: Deconstructing the corporate-state connection. *International Political Science Review, 23*(2), 159–174.

Stefansen, K., & Aarseth, H. (2011). Enriching intimacy: The role of the emotional in the 'resourcing' of middle-class children. *British Journal of Sociology of Education, 32*(3), 389–405.

van Zanten, A. (2009). Competitive arenas and schools' logics of action: A European comparison. *Compare, 39*(1), 85–98.

van Zanten, A. (2016). The Construction of Elites by Families ans Schools and the Upward Mobility Channels in France. *LAnnee sociologique, 66*(1), 81–114.

Vincent, C., Rollock, N., Ball, S., & Gillborn, D. (2012). Being strategic, being watchful, being determined: Black middle-class parents and schooling. *British Journal of Sociology of Education, 33*(3), 337–354.

Yemini, M., Maxwell, C., Koh, A., Tucker, K., Barrenechea, I., & Beech, J. (2020). Mobile nationalism: Parenting and articulations of belonging among globally mobile professionals. *Sociology, 54*(6), 1212–1229.

Yeoh, B. S., & Willis, K. (2005). Singaporean and British transmigrants in China and the cultural politics of 'contact zones'. *Journal of Ethnic and Migration Studies, 31*(2), 269–285.

7 Young People Discuss Travel

This chapter was written in collaboration with Maluhs Haulund Christensen.

Introduction

One group of travellers are glaring in their absence in research on travel and tourism: the voices and experiences of children are almost nowhere to be found in academic studies on travel (Poria & Timothy, 2014). While marketing researchers and the tourism industry have long been aware of the impact that children have on families' consumption patterns in general and tourism in particular (e.g. Carr, 2011; Nickerson & Jurowski, 2001), academic discussions on travel seem to be lagging behind on this account (Poria et al., 2005). Despite widespread recognition that the historical development of family structures has granted children an increasingly important role as democratic participants of the family, in academic literature, children are often treated as passive recipients of their parents' choices (Cullingford, 1995) or considered unsophisticated research respondents unable to generate insightful accounts on the subject of travel (Rhoden et al., 2016). This might explain why so few studies on children's role in family travel rely on data generated from children's own accounts (Poria & Timothy, 2014).

Theoretical Note

As we have illustrated across the other chapters in this book, children play a centre role in parents' decision-making processes and aspirations around family travel and they influence travel practices in multiple ways, and at different stages of the family life circle. Those studies that do exist on the matter have demonstrated how, especially in matters of deciding where to travel to, how to travel, and what accommodations to use, children play a crucial role in processes of family-making (Thornton et al., 1997; Wang et al., 2004). In one of the first survey studies addressing children and their travel experiences, as early as 1995, Cullingford explored the views that American children, between 7 and 11 years of age, hold about other countries in the context of tourism. He argued that children were likely to adopt

DOI: 10.4324/9781003056430-7

'packaged' understandings of travel, as conceptual constructs that involved beaches, pools, and nice weather, similar to how tourist industries market holidaying. He also noted that, while children were highly aware of the learning achieved through practical encounters while travelling, most significantly, through the cultural differences experienced, they still relied heavily on national stereotypes and prejudices towards certain people and places, thus suggesting the significant impact cultural representation of countries in tourism has on children's attitudes to the world. This supports the body of research that has been conducted on adults, emphasising the power of spatial imagination, and arguing that tourism materials construct places as desirable tourist destinations through image-generated narratives producing and sustaining the 'tourist gaze' (Morgan et al., 2007; Urry, 1990; Urry & Larsen, 2011). In a more recent study based on 'real time' travel data on Australian children's holiday preferences, Rhoden et al. (2016) found that the things children at the age of 9 and 10 years valued about vacationing with their family included being physically active and having the freedom and safety to play, and to making new friends independently. Holidays represented time spent with family and offered the children an escape from everyday routines and environments. On the contrary, children disliked bad weather and queuing in traffic, restaurants, and attractions, and several expressed how journeys were spoiled by travel sickness. More recently still, there has been an emergence of studies investigating the connection between children's virtual mobility, as a result of increased digitalisation, and their physical mobility, arguing that the two are supplementary activities connected in complex ways, and mitigated by sociodemographic factors (Konrad & Wittowsky, 2018).

While these studies offer interesting insights into how children experience travel through more conventional forms of tourism, a knowledge gap remains with regard to how children and young people from various socio-economic backgrounds might experience travel differently. In addition to this, research is yet to investigate how children understand travel from a mobility perspective that goes beyond tourism. As we have previously argued, mobility involves many types of travel, and can be conceptualised beyond mere physical movement. To Kaufmann et al. (2004) *motility* can be considered a form of capital that connects people's actual or potential spatial mobility (in our case travel) with social structures and dynamics that determine social mobility (p. 745). Thus, using Kaufmann et al.'s concept of motility, this chapter explores how Danish young people link social and physical mobility, and how perceptions and aspirations of travel might be influenced by perceived and actual access to mobility, given the young people's social, economic, and cultural family backgrounds. In doing so, this chapter responds to Poria and Timothy's (2014) call to capture the voices of children in relation to travel and explore how we can conceptualise their narratives around travel experiences and desires.

Methodological Note

The empirical foundation of this chapter consists of seven focus group discussions with young students attending the 9th grade at a public Danish secondary school. A total of 24 students participated in the discussions that were conducted in between classes and during lunch breaks. Considering that the participants were between 14 and 16 years of age, consent was given both by the school and their parents for young people to participate. Before the focus groups began, consent from the young people themselves was also sought. The specific public school was chosen due to its location in municipality with a reasonably large socio-economically diverse population, also represented in the school itself. Talking to children of varied backgrounds allowed us to explore how young people understand travel differently, and relative to their peers. Focussing on their past travel experiences and their future aspirations in a class context allowed us to benefit from the familiarity between peers, but also their diversity in terms of background. This proved a useful setting for examining how travel ideals were negotiated and evaluated among children from various socio-economic and cultural backgrounds.

Findings

The following analysis explores how young people talk about and discuss travel with their fellow classmates. To understand how travel trajectories and aspirations for mobility are shaped by family background, social positions, and relational negotiations, we apply Kaufmann et al.'s (2004) framework of motility to argue that mobility can be considered a form of capital that involves both *access* (including skills) and the *desire* to be mobile.

Access to Travel

Despite having grown up in the same geographical area and attending the same education institutions, travel patterns and experiences varied significantly across the young participants and their families. Most often these differences were directly related to their parents' economic situation and/ or legal status in Denmark. In addition to this, the parents' capacity to appropriate institutional resources from their children's school was also found to affect the young people's potentials to travel. In Kaufmann et al.'s (2004) words, these factors can be denounced as different contextual formulations determining the children's *access* to mobility. Access refers to how mobility is, among other factors, dependent on 'spatial distribution of the population and infrastructure, sedimentation of spatial policies, and socio-economic position' of an individual or group (p. 750). Differences in access help us explain why the different groups of young people had different mobility options at their disposal and thus different travel trajectories. In this section we take a closer look at what causes these differences in travel trajectories and how this potentially effects the imagined mobilities of young people.

Socio-economic Positions

Not surprisingly, family economic situations, at least in part, was found to influence the kinds of travel the young people had experienced. Though we did not ask directly about family income, all the young people were asked to describe what work their parents did and the level of education they had completed, which allowed us to make some assumptions concerning the family's economic resources. Seven of the participants had parents who held employment positions that required, as a minimum, a university-level degree. The parents' jobs ranged from consulting, through research and analytics, to teaching. All these young people had extensive travel experiences. Compared to the other students, these children had travelled both more regularly and farther than their peers. Liv's mother and stepfather, for instance, took her to Italy on holiday every summer and on skiing vacations most winters. In addition to this, she had visited several European metropolises. Liv's best friend, Mille, spent every second summer travelling around parts of France, where her family would follow the Tour de France bike race, for two to three weeks at a time. Ellen's family owned a summerhouse in Spain, located near a golfing complex, where they spent many of their holidays, and Cecilie had gone with her family on several trips across both northern and southern Europe. Andreas and Karl also travelled a lot – including trips to Turkey and across parts of southern Europe. Deniz usually travelled to his parent's country of birth in the summer holidays, located in the Middle East, but his family also travelled to other places in Europe, including Germany and Sweden.

   In comparison to these travel trajectories, Rasmus, David, Amalie, Oscar, Fatima, and Jasmin, among others, came from families with fewer economic resources and lower educational attainment. David, who came from a family where both parents were unemployed, had never flown anywhere with his parent on holiday. Likewise, Rasmus's parents had both been unemployed for a long period of time, with his mother only recently securing a new job in the centre of Copenhagen. According to Rasmus, this had significantly affected his family's travel opportunities. They rarely travelled to long-distance locations due to the associated costs of flying. In the following quote, Rasmus explains how his family's economic situation limits their travel opportunities:

> When we fly, which is very rarely because we don't have enough money, but when we did last fly, a couple of years ago, I thought it was nice to get away and be outside, because we used to travel to warmer countries. Denmark is such a f★★★ing cold place.

This was common among this group of respondents, to identify how financial resources shaped possibilities for where and how to travel – something their peers from more highly economically resourced families never mentioned.

> The reason why my family travelled to Greece was because it was cheap.
>
> Oscar

INTERVIEWER: Have you thought about where you would like to travel?
DAVID: Well, there is this trip with the school, where it would be me and a lot of other children going to Japan. It was planned last year before the summer holidays, but was cancelled because of the corona virus. If it gets offered again, I would like to go there. It is also ... It is a lot of money, it cost 10,000 krone to go there, so ...

They would return to the matter of money and price during many discussions, as a way of explaining or justifying their limited travel experiences. In Rasmus's view, not being able to fly robbed him and his family of the chance to travel to far-away destinations, or as he called them, 'warmer countries', that were believed to fulfil the requirements of a 'nice' holiday (Oscar). Rasmus and his family mainly spent their holidays in the neighbouring countries, Sweden or Germany. David, too, tended to highlight the price of different travel activities when talking about his own and others' experiences with travel. David had only holidayed outside of Denmark once, and this was when his grandmother gifted him a trip to Thailand as part of his confirmation present. Despite a family's economic situation restricting Rasmus from accessing certain, aspired-to forms of travel, Rasmus and his family had found meaningful ways of navigating their opportunities for being mobile:

> I know that I'll probably be going back to Sweden [laughs]. But I would like to go to Germany again. We went to Hansa Park [large amusement park] two years in a row, and then we were forced to stop. I mean Hansa Park is cheaper than Tivoli [iconic amusement park in Copenhagen], so ... why not. But the trip there also has a price [Andreas agrees with him]. But the entrance ... I mean ... the entrance and the trip are cheaper I would say.

In this short monologue, Rasmus is discussing the costs involved in visiting an amusement park in Germany. He compares it to going to Tivoli in Copenhagen and concludes that it should be feasible for him and his family to continue their routine of going to Germany. Though his family's lack of economic resources did, indeed, limit his and his family's mobility in certain ways, Rasmus sought to present his family's travel choices as having value, despite them not measuring up to perhaps more popular notions of what could be considered a desirable holiday destination (such as a place with warm weather). Using Kaufmann's notion of *skills*, it could be argued that Rasmus's family's economic situation had instilled in him certain skills and forms of knowledge, which he mobilised to evaluate and make informed decisions about which possible mobility choices to make for himself. In this way, he used his specific knowledge about amusement parks and skills of economic reasoning to make the best decision, given his mobility potentials.

Legal–Political Status

Aside from differences in socio-economic access to travel, the sample included children of parents with different legal statuses. Fatima's and Jasmin's parents, for example, were labour migrants who had moved to Denmark from the Middle East some years previously. Fatima, whose parents had immigrated to Denmark when they had been in their late 20s, had struggled to find permanent employment because they did not speak Danish. However, she travelled back regularly to her parents' home country in the Middle East. The purpose of these travels was mainly to see family and maintain contact with their friends (similar to our immigrant Israeli Jewish working-class respondents in Chapter 5). Although Fatima's parents struggled to remain economically afloat, this form of travel was affordable because they had their own house to stay in when visiting with the family. Jasmin, too, travelled to her father's home country in the Middle East every year. Her mother worked as a hairdresser and her father in a restaurant. Despite the annual visits to the father's family, Jasmin had not travelled outside of Denmark with her family. Part of the explanation for this was the fact that her mother did not have a Danish passport, which restricted her access to enter several countries.

JASMIN: There are a lot of places I want to go with my family, but my mum can't travel because she doesn't own a Danish passport.
INTERVIEWER: Did you want to travel to other places, then?
JASMIN: I really want to travel to Morocco and Turkey, but my mum can't travel there.

In this excerpt Jasmin expresses an urge to travel outside Denmark, while at the same time recognising the immobility her mother's legal position causes for her own opportunities to be mobile. Jasmin's and Fatima's accounts illustrate how motility is not only related to having the socio-economic resources necessary to travel but is also intimately linked with juridical and political conditions and structures of access (Harpaz, 2019). This demonstrates Kaufmann et al.'s (2004) point that "spatial constraints" impose differentiated potentials for mobility on different people given the context they live in (p. 752). Nevertheless, we see how immigrant background does make certain modes of travel and travel destinations affordable (see also our discussion in Chapter 5).

Institutional Networks

Despite their differences in socio-economic backgrounds, Andreas and David shared a similar interest: gaming. They spent much of their leisure time playing games on their computers. Through their shared interest in gaming and the technology that made this possible, they had both cultivated a deep interest in Japan and Japanese culture. Both of them expressed a wish to travel to and spend time in Japan. Andreas had even enrolled in Japanese language and culture classes at a local evening school. When asked where this interest in Japan came from, he said,

> I just really like the culture of Japan. It is probably one of the cultures I like most in the world. Besides our own, it [the Danish culture] is also alright. [RASMUS: 'Weeb'] No, it doesn't have anything to do with that. It is the opposite, because Weeb means that I don't respect their culture, and I do!

The two boys' fascination with Japan and Japanese culture illustrates how travelling to foreign destinations need not necessarily involve the physical movement of bodies in order for spatial connections to emerge. By virtue of the World Wide Web, the boys had built a genuine interest in a nation, its people, and ways of living. Given the relative popularity of gaming and other types of Japanese cultural contributions (e.g. Manga cartoons), the school had decided to organise a study trip to Japan in 2020. Unfortunately, the COVID pandemic meant the school trip was cancelled, but the fact that the trip could and would have been a reality for both boys highlights the potential role schools as local institutions can play in linking aspirations of mobility with actual possibilities of access. In the case of Andreas and David their school thus represented an opportunity for levelling out those socio-economic conditions that otherwise underpinned their unequal mobility opportunities. Their fascination with Japan, and the possibility that despite their different family resources, the school might organise a trip there, meant that both young men could attempt to accrue actual or aspired-to mobility experience. Especially for David, who had very little international travel experience due to his family circumstances, his aspirations for mobility seemed coherent when set alongside his ambitious future plans – to study psychology or law at university. As for our working-class families in Chapter 5, there could be argued to be a connection between social mobility aspirations through education (getting a university degree) and the desire to travel abroad (even if David's family has not been able to realise this aspiration for him). Critically too, this fascination with another culture and country through virtual mobile experiences, and a strong commitment to fulfilling their wish to spend time in Japan also appeared to shape their sense of self. We return to such self-positionings later in this chapter.

Linking Physical and Social Mobility through Future Aspirations

Being mobile through travel was both a present-day activity for many, as well as one that was tightly bound up with articulations about aspired-to futures. This gives us an opportunity to conceptualise how physical and social mobility are discursively intertwined, but also how the scope of our horizons is informed by where we are currently standing. Liv and Mille, for instance, both aspired to travel in the future – for education and work, and for leisure. Liv was planning to take a gap year before starting her upper secondary education (Gymnasium) and go to a specialist boarding school (*efterskole*). This required of Liv to leave her home, friends, and community behind and move to another part of Denmark. She said,

> I just know that I shouldn't go straight to High School right after 9th grade. It is like too much, I think. I want to go to the boarding school where I can do something with music, make some new friends, and get out of Krydsbæk [the neighbourhood where she lives] [laughing].

Alongside these aspirations to be mobile for education, Liv also dreamed about travelling to New York and the United States with her friend, Mille. The way Liv spoke of it, travel – whether it was long-distance travel to the United States or shorter distances to a boarding school in Denmark – opened up a field of opportunities she was eager to explore and gain from. It was almost as if the movement out of her current, localised life through the process of choosing to take a gap year would be enough to open up an entire field of possibilities in terms of travel, and where she might study and live in the future. The possibilities she aspired to were rather open-ended and involved broader desires of seeing other places, enjoying herself, and making new friends, which were also cultivated for her by her parents and her environment (Lareau, 2003). Fatima, on the other hand, who came from a quite different background (low economic resources and migrant background) imagined a much more concrete trajectory for her future, one that would be academically ambitious and, she anticipated, would ensure her social mobility. When talking about her future, she sketched a specific educational route, through which she hoped to one day become a social worker:

> If it is possible, I would like to take the HF [specific vocational higher education route] and become a social worker. And if this isn't possible, I would like to take a SOSU [a different vocational route, with slightly lower education requirements] and become a kindergarten educator.

None of the other young people from working-class backgrounds expressed such clearly developed plans for their futures. Drawing on Kaufmann's conceptualisation of motility, it could be suggested that Fatima's regular travel back to her parents' home country (a long-distance travel destination despite having few economic resources) and her general desire not to stay immobile by "staring out the window" in Denmark but travelling more in the future, facilitates Fatima's ambitions for social mobility. While Fatima's few but regular family holiday travels abroad might open up a field of possibilities for social mobility, Liv's much more privileged background in terms of economic resources and travel experiences ensure her field of possibilities in terms of social and physical mobility was wide open and appeared, at least in her view, completely unconstrained.

So far, we have shown how travel is related to multiple structures of access. Socio-economic resources, political-juridical conditions, and capacities to mobilise institutional opportunities all play a role in shaping young people's field of possibilities to travel and to be mobile in the present and future. This has led us to suggest that physical movement (or travel) can be interchanged with other forms of mobility. Critically, our analysis suggests we need a much more nuanced understanding of how the various elements work together – as least access does not necessarily always imply immobility or most access unlimited mobility. What our analysis illuminates is that desires for travel and/or for social mobility appear to be drawn in to make more concrete plans to be both physically and socially mobile.

Desires to Travel

All the students in this study had gone travelling with their families at least once over the course of the past three years. To most of them, travel represented a memory they looked back at and looked forward to, with excitement. In this section we look at how norms around travel were negotiated among the students. In doing so, we seek to understand what symbolic value travel has to young people, and how individual desires for travel are shaped by relative and relational social processes. This section, then, explores how certain forms of travel are deemed appropriate, valuable, and worthy of aspiration, which, in Kaufmann et al.'s (2004) terminology, is related to why people *appropriate* mobility.

Negotiating Travel Desires to Determine What Has Value

When asked why they liked travelling with their families, most of the young participants said that travel offered unique experiences, opened up particular ways of understanding the world, and appeared actively integrated into their future aspirations. Andreas described why he liked travel, noting, as a point of departure, his appreciation of learning about cultural difference:

> Denmark is a very unusual country when compared to other countries –
> also in a cultural sense. It is very fun to see other countries and how they –

at least the warm countries close to equator or even Italy – how they all sleep in their houses during the middle of the day.

Thus, for Andreas, travelling to other countries gave him the opportunity to experience cultural differences which appeared to characterise places different to his home nation. This mirrors Cullingford's (1995) finding from his study on American children, suggesting that cultural differences in dress, manners, food, and language often make the greatest impression on children and young people when going travelling. It is a way for them to not only explore other ways of living but also become aware of the 'unusual' traits of one's own culture, as Andreas formulates it. Cullingford's study also emphasised how cultural evaluations are shaped by stereotypes and prejudices around places and people that mirror larger societal discourses and representations of specific countries created by the tourism industry. In our talks with Danish young people, the United States appeared repeatedly in discussions about desirable destinations for travel. In the following excerpt, Liv and Mille explain their fascination with the United States:

MILLE: We have agreed to go together to New York, and perhaps visit the rest of the United States as well.
LIV: In the United States!
MILLE: It is a bit further away, but I think it is something completely different than Europe.
LIV: And you also hear a lot of things about the United States and there are so many exciting things to do there. It is a beautiful landscape … It is so far away.
MILLE: You see the United States all the time in the movies, so you just want to go and experience it.

Here, distance away and the saturation of American popular culture across the world have shaped Mille and Liv's desire to visit the country. Liv and Millie associated travel with adventure, seeing awe-inspiring landscapes and experiencing a culture which they, despite its distance in geographical terms, also felt they had some familiarity with. Similarly, in another peer group, Oscar, Cecilie, Deniz, and Amalie discussed their wish to visit the United States:

OSCAR: If a normal Danish person, who has never been to the United States, is asked about it, they usually say, "I really want to go to the United States".
CECILIE: Yes, also because everything is bigger.
OSCAR: Exactly.
DENIZ: I also think it is a lot nicer in the United States.
AMALIE: Yes.

OSCAR: I have also always wanted to explore the cities in the United States. It has always been something special if you see what I mean?

In this discussion, Oscar defines the United States as the "normal" desirable travel destination of Danish people, hereby setting it up as the norm for all Danish people in very few words. Besides from setting the direction for the responses that follow, Oscar's statement also shows an awareness of, and concern with, what the larger Danish population deems to be desirable travel destinations. When asked why she would want to travel to the United States, Amalie responded that she had 'always thought it looked nice in photos, videos and stuff like that, and also in movies'. As none of the four participants in this conversation have ever visited the United States, their admiration with the country is solely the product of mediated representations and shows the susceptibility of young people with regard to such representations. This affirms the body of literature arguing that tourism advertisement, and particularly efforts of 'nation branding', influence travellers' desires concerning holiday destinations (e.g. Butterfield et al., 1998; Fullerton et al., 2013).

Comparing Travel Patterns – Leading to Processes of Inclusion/ Exclusion and Identity-Making

That standards and expectations around travel were determined through social processes of comparison was evident throughout the discussions with the young people. It was most evident when the students talked about their own experiences of travel, as they tended to evaluate their own trajectories in relative terms, comparing the frequency and content of holidays with their peers. David, for example, said that he 'only' travelled with his family once a year, which he considered rare compared to his classmates, and that even the purpose of his family travels appeared to be different. While his family focussed on doing things together, other families seemed to prioritise visiting places and well-known monuments. David said,

> I think the family travel experiences I have had often set me apart from the others [friends and classmates]. Because when I hear about people travelling, they travel more than I do. So, they often travel in the summer or winter holidays and some of them also in the fall and easter breaks. Or they go on extended weekend trips of about 3–4 days. Whereas I normally travel one or two weeks only in the summer holidays. And normally, when I tell people about my holidays, I talk about what we as a family did together, like going to the beach or for a walk. And when other people talk about their holidays, they talk about the trips they went on to see, like a giant tower or … going to New York and seeing the Statue of Liberty for example.

David's account illustrates how travelling is a differentiated practice, subject to relative scrutiny and reflection. Travel patterns work to signal differences in families' resources and access to travel. The regular sharing of family travel experiences in everyday conversations and during class discussions had made differences in travel related to where, how, and why families travelled visible. Although it was never said directly, there was often a sense during the discussions in the groups that for the children who travelled less, their own way of travel was associated with feelings of doubt or insecurity. Amalie, for example, remained silent throughout most of the discussions with her classmates. Having 'only' travelled to Germany with her family during the past three years, she had travelled the least among the classmates in her discussion group. In the group discussion she remained silent for as long as possible and then, when asked directly, said that they primarily travelled there to shop, and they did not really do anything else:

INTERVIEWER: [...] What did you do when you were travelling with your families?
CECILIE: I have also been on a skiing trip and beach holidays.
DENIZ: Yes, me too, I have also been on beach holidays. Or exploring the city where we go on trips and stuff like that.
JASMIN: Yes, me too.
[AMALIE IS QUIET]
INTERVIEWER: What about you [Amalie]?
AMALIE: Just went shopping and stuff like that. We didn't do that much.

Amalie's silence could be connected to her awareness that she is unable to share similar experiences of travel. Her use of the words 'just' and 'didn't do that much' work to downplay the impact of the experiences she has had when travelling with her family. Whereas Cecilie and Deniz speak excitedly about the multiple holiday forms they have been on, Amalie appeared to dismiss the influence of those few travel experiences that she had had. Thus, it might be said that her lack of stories to share meant she was positioned as having nothing of value to contribute within this particular discussion.

The strong symbolic value that travelling held for most of the students meant that frequency and forms of travel became elements for comparison between experiences, that worked, as we saw above, to create social inclusion and exclusion among the pupils. This could, in turn, affect young people's understandings of themselves and their families. This was not only true for those students, like David and Amalie whose family travel practices did not meet the standards of regular and long-distance travel. Liv, for example, used her taken-for-granted ability to be mobile as a way of distinguishing herself from others:

None of my friends have travelled to Iceland, I think. And when I am in Italy – there are a lot [of my friends] who have been there, but I think like ... When we are there, we are there in another way sometimes. Specially because we need to pick up wine and stuff like that [...] Like, we also know people who live there, some of our closest friends who we often go to see, we have been there a lot of times because they have a vineyard and so on. Because they are Danish, right? [...] So, I don't think that a lot of my friends do it [that way], but it is a nice way of doing it.

Just like Liv, Ellen was also aware that her family's access to travel exceeded that of most of her classmates: 'Well, we have a summerhouse down there [in Spain]. It doesn't seem like most people have a summerhouse in other countries than Denmark. Probably not [...]'. Meanwhile, Rasmus had had much more limited opportunities for travel since his parents were long-term unemployed. Despite not flying abroad anymore, Rasmus's family still found ways to travel:

> I don't think I have flown for like four years [...] We do go to Sweden at least three times a year. To our summerhouse. Then, we have also been to Germany for the past two years – not this year due to the coronavirus, but we usually go to the amusement parks in Germany because they are just way better. That is really it. We don't really get out [of Denmark] otherwise.

Compared to his peers' stories of holidaying, Rasmus felt the need to end his contribution by qualifying his experiences as 'really get[ting] out' of Denmark. Though, as argued earlier, Rasmus worked hard to create a sense of his parents making the most of the limited resources they had, ultimately, he was very aware that his mobility history held less value when compared to those of his peers.

Despite the variation in financial resources, most of the young people understood family travel as a valuable activity, and used it – consciously or unconsciously – as a practice of distinction that worked to re-affirm one's inclusion in a peer group where travel abroad is taken for granted (as Rasmus worked hard to claim when recounting the travel he did with his family). Rather than being a fixed set of prerequisites, desirable travel forms and experiences are negotiated through social interaction and discussions, through which young people become aware of their own practices and learn to assess opportunities in accordance with social standards of travel mobility.

Rejecting Travel Norms

Interestingly, not all the students expressed consistent opinions about travel as an exclusively positive practice, and none talked about environmental concerns in relation to air travel (as we discuss in detail in Chapter 9). Despite voicing initial excitement about the idea of travel, Oscar and Amalie both changed their opinion about family travel somewhere along the way during the group discussion. Amalie had previously confided that she dreamed of travelling to Mexico, Greece, and the United States, and Oscar had been among those who took the United States as an obvious, desirable travel destination. When asked why they had not travelled as much as some of their peers, they responded as follows:

OSCAR: I have never really travelled so much. I have always been someone who prefers to just stay at home.
AMALIE: Yes, I like that better too. Rather do something fun here [at home]. [...]
INTERVIEWER: How come you [Oscar and Amalie] prefer to stay at home [in Denmark]?
OSCAR: I think, the thing that makes me not want to go away is, you know... I'm not one for new places.
AMALIE: Exactly, yes.
OSCAR: It is mostly that [...] [I prefer] always doing things the same way, right? The thing about going to new places, I have never liked that. That is the reason why every time we [I and my family] travelled to Spain it was almost always the same location we went to.

In this excerpt Oscar and Amalie come to an agreement about the idea that staying at home can be preferable to travelling abroad. They hereby reject the norm that travel is a universal good and something to aspire to. Their accounts suggest that travel to them is not necessarily related to opening up a broader field of opportunities. Oscar explains his disinterest in travel with reference to how he is 'not one for new places', and Amalie, who is less vocal, supports him. While an obvious answer to the question being asked by the interviewer would have been to say that both of their parents did not have the financial resources to travel often or long-distance, Oscar worked hard in his response to mobilise an image of agency, where his lack of travel experiences are attributed to a justified and strongly held personal disinterest, and not to the material scarcity of his family. Later in the discussion, Oscar returns to this position-taking, rejecting the excitement or benefit of new experiences gained through travel abroad:

When I was younger, we went to Egypt. I have always had problems with stuff like food, because there is such a big difference between food [in different countries], right? So ... I think it is a little difficult to travel to another country where there aren't the same things, right?

To back up this experience, he used the example that milk and water usually tastes bad abroad, when compared to that in Denmark. Eventually he argues that his discomfort with new experiences and other ways of living is the reason why he and his family have almost always travelled to the same location. Furthermore, in the group discussion, Oscar appeared not content to justify his own experiences, but would actively comment on other's experiences and desires. For example, when Amalie was asked to explain why she wanted to go to Mexico, Oscar commented, 'Mexico, I have actually never ever wanted to go there. I have always felt Mexico was just [...] Mexico [...] Well, just something for itself, and then something like Spain I guess', and when Deniz mentioned, that he was keen to visit Barcelona, Oscar reported that in his opinion the things to see in Barcelona, including monuments and beaches, 'sucked'. Although Oscar's reactions towards his classmates' accounts might seem unnecessarily negative, it served an important function to shoring up his own self-image. Only by deeming travel undesirable and by favouring experiences gained from holidaying in Denmark was he able to evaluate his family's limited access to mobility in positive terms.

Like Oscar, David's parents also had restricted access to long-distance and frequent travel opportunities. As previously mentioned, he had never gone on holidays with his parents that required air travel. His family usually spent their holidays in a summerhouse in Denmark or in a neighbouring country, reachable by car or public transport. However, as a present for his confirmation, his grandmother had gifted him a trip to Thailand:

> That time I went to Thailand, that was [with] my grandmother, she owns a house in Thailand and she normally travels there four times a year as a minimum. And then she wanted me to come, also because I have never tried flying before, and she thought that that was something I had to try – that that would be the best of the best, right? Normally we go to a summerhouse instead of going travelling, because we think, that there is a lot to Denmark that we actually don't know. That that can in fact be a form of travel, even if you already live in the country. And that is a fun concept, I think.

David here actively tried to defend the idea that for a holiday to be considered travel, one need not go abroad. Despite having been provided with an opportunity to take part in, what most of his peers might perceive as 'the best of the best' of travel experiences, he defends the kind of travel practice that his parents have introduced to him. His parents' opinion has been transformed into a 'we think', that places him in opposition to the 'normal' way of understanding travel, and legitimises the way his family travels. In that sense we witness the ways in which travel desires are examined, defended, and reshaped in these multiple negotiations between young people, peers, and their families.

Conclusion

This chapter has explored how young people from different family backgrounds talk to each other about travel. We have argued that mobility, as part of (family) travel, is understood as a practice that requires resources (financial, institutional, and legal) and a practice that allows for the accrual of resources that play a role in feeling included and/or being able to distinguish oneself. This mirrors Kaufmann et al.'s (2004) words that 'movement can take many forms, that different forms of movement may be interchangeable, and that the potentiality of movement can be expressed as a form of "movement capital"' (p. 752). While this field of possibilities is already firmly in place for some, it might be emerging more slowly and along less steady routes for others. We saw how resources such as income, cultural interests, and social connections also shaped being able to activate the field of possibilities in the present, while for others – the field of possibilities was not only being articulated and aspired to but was something they hoped would become actualised further down the line. In this way, mobility is a capital one can accrue or actively seek to activate as part of the broader plan for (social) mobility. Although we do not have enough data to fully develop this argument, some of the data suggested that motility (as already accrued or as something desired) might play a role in influencing possibilities for social mobility. Finally, our analysis illustrates how actual or aspired-to mobility might play a role in shaping relations between young people in school setting, where children of diverse backgrounds meet, and how this might function as a mechanism for inclusion or exclusion, that can impact on one's identity-making project and on the sense of one's family's worth.

References

Butterfield, D. W., Deal, K. R., & Kubursi, A. A. (1998). Measuring the returns to tourism advertising. *Journal of Travel Research, 37*(1), 12–20.

Carr, N. (2011). *Children's and families' holiday experience.* Taylor & Francis.

Cullingford, C. (1995). Children's attitudes to holiday overseas. *Tourism Management, 16*, 121–127.Fullerton, J. A., Kendrick, A., & Golan, G. J. (2013). Strategic uses of mediated public diplomacy: International reactions to U.S. tourism advertising. *American Behavioral Scientist, 57*(9), 1332–1349.

Harpaz, Y. (2019). *Citizenship 2.0: dual nationality as a global asset.* Princeton University Press.

Kaufmann, V., Bergman, M. M., & Joye, D. (2004). Motility: Mobility as capital. *International Journal of Urban and Regional Research, 28*(4), 745–756.

Konrad, K., & Wittowsky, D. (2018). Virtual mobility and travel behavior of young people – Connections of two dimensions of mobility. *Research in Transportation Economics, 68*, 11–17.

Lareau, A. (2003). *Unequal childhoods.* Berkeley.

Morgan, N., Pritchard, A., & Pride, R. (2007). *Destination branding: Creating the unique destination proposition* (2nd ed.). Elsevier Butterworth-Heinemann.

Nickerson, N., & Jurowski, C. (2001). The influence of children on vacation travel patterns. *Journal of Vacation Marketing, 7*, 19–30.

Poria, Y., Atzaba-Poria, N., & Barrett, M. (2005). The relationship between children's geographical knowledge and travel experience: An exploratory study. *Tourism Geographies, 7*(4), 387–397.

Poria, Y., & Timothy, D. J. (2014). Where are the children in tourism research? *Research Notes/Annals of Tourism Research, 47*, 77–95.

Rhoden, S., Hunter-Jones, P., & Miller, A. (2016). Tourism experiences through the eyes of a child. *Annals of Leisure Research, 19*(4), 424–443.

Thornton, P. R., Shaw, G., & Williams, A. M. (1997). Tourist group holiday decision-making and behaviour: The influence of children. *Tourism Management, 18*(5), 287–297.

Urry, J. (1990). *The tourist gaze.* Sage.

Urry, J., & Larsen, J. (2011). *The tourist gaze 3.0.* Sage.

Wang, K. C., Hsieh, A. T., Yeg, Y. C., & Tsai, C. W. (2004). Who is the decision-maker: The parents or the child in group package tours? *Tourism Management, 25,* 83–194.

8 Alternative Modes of Family Travel
New Articulations of Global Citizenship Education

Introduction

Travel is explicitly and implicitly entangled in understandings of global citizenship (Goren & Yemini, 2017; Goren et al., 2020; Oxley & Morris, 2013). When asked, many families understand travel as central to enabling their children to encounter new people, cultures, and landscapes (Adams & Agbenyega, 2019), which, in turn, they hope will engender a greater sense of their common humanity and a commitment to sustaining the environment. However, not all families will necessarily articulate such 'lofty' aspirations; many would be more likely to see travel as central to promoting a desire for mobility for future educational and employment destinations, and to serve as a form of distinction within local spheres of practice (Weenink, 2008, 2009). Either way, the physical act of travel in conjunction with family discussions about where to go and what to do there should be understood as tied to often implicit conceptions of what global citizenship education (GCE) is, whether it is aspired to, and how to actualise it. While the book, in general, is concerned with short-term family travel, here we delve into two distinctive modes of travel which are less common, but significant nonetheless for understanding how travel is shaped by parental broader engagements with the world and critical issues facing us today. While in earlier chapters we draw on Andreotti and colleagues' (2015) model for global mindedness dispositions, here we focus on GCE, as a more encompassing and currently prominent concept for analysing our relationship to places and people around us.

While most of the writing in the field of GCE is focussed on teachers and formal schooling (Estellés & Fischman, 2021; Goren & Yemini, 2017; Yemini et al., 2019), we seek to highlight the potential role of parents and the global citizenship education work they perform here through the act of travel. Drawing on recent global citizenship education literature – specifically, ideas developed by Pashby and colleagues' (2020) review – we evaluate whether two particular approaches to travel – abstaining from travel or extended periods of travel as a family – constitute new methodological, epistemological, or ontological relationships to global citizenship education.

DOI: 10.4324/9781003056430-8

Theoretical Note

The purpose of 'education,' whether conducted within families or within institutions, is usually broadly conceived. Yet, increasingly, critics see education systems and family practices of education taking a rather instrumental approach (Goren & Yemini, 2017; Resnik, 2009). Thus, even as curricula take on a more 'global' dimension, this is often understood as being driven by the need for students to be able to compete for work in a global labour market (Engel et al., 2019). However, threaded through such an objective we often encounter broader aspirations for young people to also develop the skills to engage in 'global problem solving' and address particular issues that globally connected contemporary societies must tackle (Dill, 2013; Goren & Yemini, 2018; Yemini & Furstenburg, 2018). Such engagement would require an awareness of what issues affect us at a global level or are experienced in different ways across various parts of the world. Another element of education is the quest to instil empathy for people of different origins and an ability to engage in multi-cultural environments. Not only do schools and formal curricula take on the responsibility of 'teaching' students about these issues, but parents are also explicitly and implicitly concerned with these three dimensions of education. Their interests in this regard are reflected in the decisions they make about the kinds of formal and informal education they seek out for the children (Vincent & Maxwell, 2016) – the types of school choices, extra-curricular activities, decisions about whether to buy into private tutoring or not – as well as the broader kinds of family learning practices they promote. This latter sphere is where we place the practice of family travel, namely, within the broader project led by parents of educating a child. Regardless of the extent of a family's resources, decisions about where and how to travel have meaning and a purpose.

So far, in this book, our analysis has suggested that most parents connect travel to their quest to 'open up the world' for their children in various ways. Our analysis, informed by Kaufmann et al.'s (2018) development of mobility as a form of capital that they dubbed 'motility,' unsurprisingly, interpreted family travel as a neoliberal-infused practice, focussed on securing the social mobility of their children. But in all cases, travel was focussed on an engagement with the 'Other' – be it cultural, geographic, the uncomfortable unknown – and a desire for their children to embrace this. In this chapter, we seek to examine family travel practices within a novel framework – that of global citizenship education (GCE). In so doing, we explore whether we can elucidate other interpretations of the meanings of family travel for the formation of social relations on national and transnational scales. Given that GCE is now part of the United Nations Sustainable Development Goal 4, Target 7, there is a need to understand the role family travel can play in its interpretation and implementation, as well as the outcomes of such an imperative promoted broadly by society, and more specifically by the education system.

Many reviews of the scholarship on GCE exist already (Goren & Yemini, 2017; Goren et al., 2020; Oxley & Morris, 2013; Pashby et al., 2020). In the most recent and arguably innovative review of GCE frameworks/typologies, Pashby and colleagues (2020) proposed a useful approach for understanding various forms of GCE, based on several basic orientations (neoliberal, liberal, and critical) and, importantly, the interfaces between these orientations. The authors revealed that much of the contemporary practice and conceptualisation of GCE remains embedded within colonial power relations, and specifically within a neoliberal-liberal orientation. This reality, they argued, poses challenges for using GCE to think differently or recalibrate existing relations of inequality, or for 'pluralising possibilities for shared futures' (p. 150). Pashby and colleagues (2020) argued that in order to explore alternative ways of conceiving of, and practising GCE, we need to examine changes in our approach, on methodological ('doing'), epistemological ('thinking'), and ontological ('being') levels. These kinds of differentiations could be aligned to some extent with Andreotti et al.'s (2015) different levels of engagements with the 'Other' via different dispositions towards global mindedness. We find these analytical prisms useful for trying to make sense of different orientations towards family travel, how these might be linked to conceptualisations of GCE, and the way in which travel constitutes different modes of thought about our relationship to the world and its people, places, cultures, and animals. The emergence of environmental sustainability as a newer but evermore critical focus of GCE (Pashby et al., 2020) also aligns with increasing concerns about the impact of travel on climate change.

Methodological Note

This chapter focusses on two specific groups of parents, identified while we were examining for examples of a broad range of parental travel preferences. These two groups developed their own travel strategies, proposing an alternative to the traditional understanding of where, how, and why families travel. These two groups were 'middle-class' in terms of their professional and education backgrounds, but actively resisted the now normalised approaches to travel found among the middle-class groups studied.

We purposively sampled secondary data via online travel blogs (as per Banyai & Glover, 2012), media reports, as well as examining whether any families in our open-ended survey (sent out via a range of personal and academic networks) undertake different kinds of travel practices from those found in our initial interviews conducted with Israeli families. We undertook web searches in English and Hebrew using keywords (blogs, travel, families, children) (Banyai & Glover, 2012) both using Google and Facebook. We then categorised all the identified webpages (over 3,000 results) by type of travel described and types of issues related to travel identified. In so doing, we found a small minority of families who were practising two distinctive travel behaviours, which we conceived of as 'exit strategies' from the normative middle-class family practices: those who abstain from air travel entirely, and those who travel as families for extended periods of time (of at least one full year).

In what follows we first show how these two opposite travel behaviours aim to cultivate GCE, but also how GCE and the environment become linked. Next, we examine the extent to which these challenges to accepted middle-class norms about family travel constitute a generative change that open up possibilities for different approaches to global citizenship education.

Findings

The Quest to Develop GCE by Customising Travel Behaviours

Both types of families argued that their travel choices are directly motivated by their desire to influence their children's development and education. Indeed, travel or abstaining from it was a major cultivation strategy these parents employed. The question about whether to travel or not reflected their perceptions of what was good and bad in the world, and their role in mediating this. GCE was often drawn on in their narratives as guiding their decision-making, as can be seen in the following extracts.

> They are open, totally citizens of the world. They can easily create social contact with children and adults of different ages, speaking different languages. Just hanging out at the beach, learning so much from their environment.
>
> Mother of three (ages 13, 9, and 3), has been travelling for two years through India, Nepal, Cambodia, and Thailand

> We want them to become true citizens of the world, taking full responsibility of their actions. We buy smart, recycle, and make them to think of the environmental cost of our daily decisions.
>
> Father of two (ages 8 and 5), who avoids air travel

In the above two extracts, the parents argue that avoiding air travel or travelling as nomads constitute behaviours that instil global citizenship. By remaining consistent with their chosen travel behaviour, the parents raise their children to become global citizens as per their respective understandings of the concept – either by getting to know people from different backgrounds or by acting environmentally consciously.

Families travelling for long periods of time were committed to cultural immersion, but also got involved in volunteer work with marginalised populations, which a father of two abstaining from travel articulated as 'giving back to our world'. The emphasis was also on the intention, the disposition to be a global citizen.

Trying. Not being afraid of trying. Trying to build international relationships, trying to travel with a small baby, trying to travel with another
small baby, hitch-hiking on boats at the end of the world with two of
them, trying to help as many refugees in Berlin as possible, trying to talk
to those who are very far from us, trying to do something by organising
a Civil March for Aleppo.

Travelling Family blog

Yet, these exit strategies from mainstream middle-class family travel also presented some challenges, as for example, families could not follow cultural traditions and stay connected to family and friends as easily, as explained below.

It's Passover eve, and we are taking advantage of the time to send and
receive greetings from family and friends. Some Israelis delay their travels
and stay in Pokhara and Kathmandu to celebrate the Passover night in
Chabad. We haven't even considered it. We are definitely cut off from
the Jewish experience in the past year and do not make an effort to connect with it. It's strange to follow here, via WhatsApp, the Passover night
preparations in Israel and elsewhere.

Mother of two (ages 11 and 9), has been travelling for
over a year in India and surrounding countries

Meanwhile, other the families mentioned the (critical) questioning they
faced, and the felt need to explain and justify their actions.

We are not lunatics; we do get new stuff when there is a real need and
we do consider taking a boat to travel abroad. The world is endless
and we want to experience it … Perhaps they have not been sight-
seeing in classic Europe, but the skills they gain here are super important. They know everything about the climate change and they will
be equipped well [for the future].

Statement made on the Zero Waste Forum on Facebook
by a father whose family refrains from air travel

Alongside a keen commitment to experiencing different cultures and geographical diversity, and aiming to become a global citizen through mobility,
the overwhelming and larger and more vocal group were those who were
actively abstaining from travel, because of their commitment to environmental sustainability, which they understood as the most critical articulation of
global citizenship education.

GCE and Environmental Sustainability

Parents linked their travel choices to environmental concerns, and in turn
these were associated with a form of GCE. Families, across both groups

in fact, aimed to become less consumerist, to recycle, and to live more sustainably.

For example, a father of two travelling with his family in Nepal and India explained how his family decided to change the foci of their lives:

> You don't need all your stuff here. It is 15 kg maximum that your porter can carry and that's it.

Or as a mother of two, who abstained from travel, said in her blog entry,

> Yes, we have our monthly budget and we just don't spend more, under any circumstances. It all actually started ten years ago when we suddenly found ourselves unemployed, after earning and spending so much money while working in the high-tech industry. But now, we are very stable financially, but decided to live this way to save the planet, adding our share of effort.

Another father of a family that abstains from travel said,

> We are not stingy as people might think, we have a nice apartment and we do things we love; we are just not taking part in this race of consumption.

Furthermore, parents shared with us how they performed their 'environmental calculations'. Thus, for example, a mother of three explained their environmentally based decision to travel for two to three years through India and Thailand as follows: 'In terms of being environmentally friendly, it is much better to take these long and sometimes expensive flights for a long stay instead of going on a short vacation'. Meanwhile, *The Guardian* (a British newspaper) ran an article on families' environmental reasoning around travel, described the decision-making process of one couple:

> Elisa Vertue, 41, lives in south-east London with husband James Wilson, 43, and children Maia, five, and Rudy, four, but is originally from Milan and takes regular trips home. The family travel back to Italy at least twice a year, sometimes flying and at other times taking their motorhome. But Elisa and James, who run their own IT support business, say they are increasingly thinking about the impact on the environment. In July this year the family is going to northern Croatia to meet Elisa's sister and her family for a three-week holiday. Elisa had looked into flying and the price was reasonable for the short hop to Europe, but instead the family has decided to take their motorhome and will make two stops in Europe on both legs of the journey. "When I thought about it, I didn't want to fly," says Elisa. "If we take the motorhome there will be less carbon emissions. We can also make the journey fun for the children. They will see

a lot more of Europe as we will make stops in France and Austria." Elisa paid £123 for the Dover to Calais ferry crossing and £145 for the Euro-tunnel on the return journey … Elisa estimates it will take between two and a half and three full tanks of petrol to drive the 17 hours to Vrsar in Croatia. This could cost more than £250 for the 2.8 litre turbocharged motorhome, but they won't need to pay for hotels on the journey or once they arrive. "As a total holiday package, taking the motorhome works out cheaper as we won't have accommodation costs once we get to Croatia," says Elisa. "But our carbon footprint should also be lower as we've chosen not to fly."

Finally, a website dedicated to different types of family travel described the perceptions of a mother who took her six-year-old daughter on a mothers-and-daughters vacation that was designed for families who wanted to combine travelling, interactions with cultural 'Others' and with a strong environmental focus:

> No work. No cell phone. No distractions. I wanted an experience that included a cultural immersion program to give her a glimpse into a world outside of her own, adventure to keep her engaged, outdoor time to keep us active, and relaxing time to allow us to recharge together. As a sustainability expert, it was also important to me that we go with a travel company that puts giving back to the environment, local economy, and community above profits.

Thus, exit strategies from expected middle-class travel practices did differ among these two 'outlier' groups (some were mobile internationally, other weren't or did not travel as far), but they were driven either by a commitment to cultural immersion in new spaces and environmental sustainability.

Discussion

Travel is part of GCE, but it remains a controversial dimension. Indeed, to be part of the world, one is expected to get to know the world; yet travel is wrought with tensions because it involves environmental damage either through carbon emissions through air travel, the ruining of untouched nature to build up the tourism industry, and other negative consequences such as transnational mobility of low-paid staff to work in large tourism complexes, the use of tradition lands without proper consent or recompense, and so forth. We argue that parents in the two groups presented here seek to actively challenge common middle-class travel practices, choosing distinctive and arguably oppositional ways to practice family travel. While GCE implicitly includes the need to travel, these families are re-imagining travel, by increasing the range of possibilities or what is intelligible about how we can perform global citizen education through parenting and travel. These parents featured here are creating new narrative possibilities for what travel can look like and why it is necessary and worthwhile. They argued that they perceived other families as 'framed by a limited range of possibilities, and thus closed off from imagining viable alternatives' (Pashby et al., 2020, p. 158; (see also the narratives of parents in Moltz, 2021 and her study on 'world-schooling' where education is undertaken through long-term travelling, who make similar arguments).

Pashby and colleagues (2020) identified three different ways to understand the practice of GCE: methodological, epistemological, and ontological. Based on our larger study, we argue that the normative discourses driving GCE among middle-class and globally mobile middle-class families are largely methodological – with family travel acting as enabler for various aspired-to future trajectories. However, the families who are the focus of the present chapter, also middle-class, arguably go beyond the methodological approach that links GCE with ideological concerns, which in turn drives travel behaviours. These families directly ask questions about injustice, sustainability and worth, and find ways to practise their choices, often in the face of criticism or puzzlement from family and friends. Thus, it is more than an epistemological approach to family travel. As evident from the quotes shared above, these families appear to be living and breathing their approach to GCE, even if it is uncomfortable, unusual or limiting. They do not just question family travel practice, they dare to act differently.

The question is how fundamentally such position-taking around travel – with a view to a 'visiting' global mindedness disposition (Andreotti et al., 2015) as many of our 'nomadic' families express, or putting our relationship to, and responsibility towards the environment as the guiding principle governing family travel practices – could be the beginning of an ontological change in our relationships to global citizenship work done in families. To examine this more closely one would need to investigate everyday practices, as well as travel practices, in our view. Also, some scholars in the field of GCE would argue that an ontological approach can only occur within a postcolonial framework, which privileged people from the North cannot fully inculcate into their practices due to their own positionality.

The chapter makes two arguments. First, we made the connection between travel and GCE as a practice that occurs within families, and as an important dimensional of concerted cultivation work (Lareau, 2003). Second, we explored the link between GCE and environment-related education focussing on families and their engagement with these intersecting principles through travel. We propose therefore to scholars and practitioners working with GCE, that seeking to critically engage with the concept of travel – its purposes, outcomes, and responsibilities – may be one lever through which GCE could bring into discussion the many pressing environmental concerns we face today, while also keeping connected to another critical foundation of GCE – engagement with the 'Other'.

Acknowledgment

This chapter draws on data and arguments already published in Yemini and Maxwell (2021). Alternative modes of family travel: middle-class parental 'exit' strategies as a different orientation towards global citizenship education. *Globalisation, Societies & Education*, online 8 February.

References

Adams, M., & Agbenyega, J. (2019). Futurescaping: School choice of internationally mobile global middle class families temporarily residing in Malaysia. *Discourse: Studies in the Cultural Politics of Education*, *40*(5), 647–665.

Andreotti, V., Biesta, G., & Ahenakew, C. (2015). Between the nation and the globe: Education for global mindedness in Finland. *Globalisation, Societies and Education*, *13*(2), 246–259.

Banyai, M., & Glover, T. D. (2012). Evaluating research methods on travel blogs. *Journal of Travel Research*, *51*(3), 267–277.

Dill, J. S. (2013). *The longings and limits of global citizenship education: The moral pedagogy of schooling in a cosmopolitan age*. Routledge.

Engel, L. C., Rutkowski, D., & Thompson, G. (2019). Toward an international measure of global competence? A critical look at the PISA 2018 framework. *Globalisation, Societies and Education*, *17*(2), 117–131.

Estellés, M., & Fischman, G. E. (2021). Who needs global citizenship education? A review of the literature on teacher education. *Journal of Teacher Education*, *72*(2), 223–236.

Goren, H., & Yemini, M. (2017). Global citizenship education redefined–A systematic review of empirical studies on global citizenship education. *International Journal of Educational Research*, *82*, 170–183.

Goren, H., & Yemini, M. (2018). Obstacles and opportunities for global citizenship education under intractable conflict: The case of Israel. *Compare: A Journal of Comparative and International Education*, *48*(3), 397–413.

Goren, H., Yemini, M., Maxwell, C., & Blumenfeld-Lieberthal, E. (2020). Terminological 'communities': A conceptual mapping of scholarship identified with education's 'global turn'. *Review of Research in Education*, *44*(1), 36–63.

Kaufmann, V., Dubois, Y., & Ravalet, E. (2018). Measuring and typifying mobility using motility. *Applied Mobilities*, *3*(2), 198–213.

Lareau, A. (2003). *Unequal childhoods*. Berkeley.

Oxley, L., & Morris, P. (2013). Global citizenship: A typology for distinguishing its multiple conceptions. *British Journal of Educational Studies*, *61*(3), 301–325.

Pashby, K., da Costa, M., Stein, S., & Andreotti, V. (2020). A meta-review of typologies of global citizenship education. *Comparative Education*, *56*(2), 144–164.

Resnik, J. (2009). Multicultural education–good for business but not for the state? The IB curriculum and global capitalism. *British Journal of Educational Studies*, *57*(3), 217–244.

Vincent, C., & Maxwell, C. (2016). Parenting priorities and pressures: Furthering understanding of 'concerted cultivation'. *Discourse: Studies in the Cultural Politics of Education*, *37*(2), 269–281.

Weenink, D. (2008). Cosmopolitanism as a form of capital: Parents preparing their children for a globalizing world. *Sociology, 42*(6), 1089–1106.

Weenink, D. (2009). Creating a niche in the education market: The rise of internationalised secondary education in the Netherlands. *Journal of Education Policy, 24*(4), 495–511.

Yemini, M., & Furstenburg, S. (2018). Students' perceptions of global citizenship at a local and an international school in Israel. *Cambridge Journal of Education, 48*(6), 715–733.

Yemini, M., & Maxwell, C. (2021). Alternative modes of family travel: middle-class parental 'exit' strategies as a different orientation towards global citizenship education. *Globalisation, Societies and Education*, 1–12. Ahead of print.

Yemini, M., Tibbitts, F., & Goren, H. (2019). Trends and caveats: Review of literature on global citizenship education in teacher training. *Teaching and Teacher Education: An International Journal of Research and Studies, 77*(1), 77–89.

9 From Tourist Gaze to Carbon Gaze

Travelling in a Time of Climate Concerns

Introduction

When Urry published his seminal work, *The Tourist Gaze*, in 1990, there was little to no premonition of the role that environmental and climate-related concerns would come to have on travel. Primarily preoccupied with the historical and societal preconditions that enabled the modern organisation of mass tourism and the cultural production of aesthetic tourist pleasures, Urry (at least in this iteration of his work) paid little attention to the climate-related costs of those transportation technologies and modern tourism industries. These have now become objects of critical attention in what has been named the era of anthropocentric climate changes (Zalasiewicz et al., 2010), with which Urry has also engaged in later work with Larsen (2011).

Over the last years, changes to the climate due to industrial, consumerist, and transportation related practices have become a major social concern, under which human conduct in all its variants is being questioned – including that of air travel (see, e.g. Garnaut, 2011; Hache & Latour, 2010; Higham et al., 2013; IPCC, 2013; Urry & Larsen, 2011). Tourism accounts for 5% of global carbon dioxide emissions (Peeters & Dubois, 2010), 40% of which is attributed to aviation (Gössling, 2009). However, given the current rate of growth of the aviation industry (COVID-19 notwithstanding) in both actual and relative numbers compared to other industries (Mayor & Tol, 2010) and the predicted increases in aviation demands by a growing global middle-class population, projections suggest that these numbers will rise to 40% of total global CO_2 emissions by 2050 (Dubois & Ceron, 2006; Gössling & Peeters, 2007).

Numerous political attempts have sought to respond to the world's growing carbon footprint. These include discourses of 'green growth' and 'sustainable development' that rely heavily on technological innovation and debunk the notion that reductions in carbon emissions are incompatible with continued economic growth (Irwin, 2001). Such neoliberal discourses are also visible in policies around aviation-caused pollution, where national and international government bodies have failed to enforce political and economic regulation on the international flight industry (Eickhout & Taylor, 2016). Instead, an increasing attention has been given to the importance of effectively changing

DOI: 10.4324/9781003056430-9

populations' social practices and patterns of consumption (Shove et al., 2012). Scholars continue to debate the existence, potentials, and sovereignty of *political* or *ethical consumers* (see Carfagna et al., 2014; Jacobsen & Dulsrud, 2007; Latour, 2014). Yet many scholars, politicians, businesses, and social movements alike agree that mainstream approaches to mitigating carbon emissions from air travel depends on the abilities of powerholders to facilitate the emergence of voluntary consumer behaviour, that is, to encourage the public to live more carbon neutral lifestyles (Higham et al., 2016). As Shove (2010) has argued, the overwhelming focus on private consumer behaviour, and lifestyle changes in large governmental research programmes[1] indicate that "issues of climate change have been framed in terms of an already well-established language of individual behaviour and personal responsibility" (Shove, 2010, p. 1274). The moral brunt of reducing the enormous CO_2 emissions caused by air travel has, in other words, been left to the individual consumer (Barr et al., 2011; Higham et al., 2019).

In this chapter, we investigate the personal and social consequences of a consumer-driven discourse on climate change mitigation. To do so, and building further on the 'unusual', non-normative patterns of travel discussed in Chapter 8, we explore what might be considered the most recent example of a climate-related stance on political consumption: *flight shame*. Translated from the Swedish word *flygskam*, flight shame was originally coined by Swedish media in 2018 to denote a tendency among environmental movements, in particular Fridays For Future, where the need to reduce greenhouse gas emissions has been framed as an issue of personal responsibility (Gössling et al., 2020). It has been defined as 'an individual's uneasiness over engaging in consumption that is energy-intense and climatically problematic'. Since 2018 the concept has been rapidly adopted by the media across the world, where it has gained increasing resonance with populations outside Scandinavia (Gössling et al., 2020). In the following section we give a brief introduction to our theoretical framework, through which this chapter explores the concept flight shame and its significance to practices of family travel.

Theoretical Note

Despite detectable increases in awareness and concerns about climate-related impacts of human consumption (Higham et al., 2016; Randles & Mander, 2009), studies suggest that behavioural changes in the form of actual reduction of air travel have still not occurred (Alcock et al., 2017; Cohen et al., 2018; Hares et al., 2010). This has led numerous researchers to conclude that we are dealing with a *value-action gap*, understood as a discord between awareness and attitudes, and actual behavioural change and consumption (see Hares et al., 2010; Hibbert et al., 2013; Kollmuss & Agyeman, 2002). While an overwhelming amount of time and research has gone into solving this puzzle of the value-action gap, scholars of *social practice theory* have claimed that 'the gap is only mystifying if we suppose that values do (or should) translate into action' (Shove, 2010, p. 1276).

Social practice theory draws on a Bourdieusian conception of social practices and rejects the image of rational agents making deliberate behavioural choices based on their individual values (Warde, 2014), that most research on climate-friendly behaviour tends to rely on. To understand how practices around consumption or behaviour arise and evolve, social practice theorists focus on how 'perceptions, interpretations and actions' are shaped by social and collective relations and structures (Hargreaves, 2011, p. 79). As such, behaviours are considered to be negotiated both within structures of provision (meaning infrastructures, financial means, etc.) and social structures (such as peer groups, networks, and institutions) (Shove, 2010). Such negotiations involve both notions of social identities and their position within a larger social field (Hibbert et al., 2013) as well as symbolic and affective meanings of objects and forms of behaviour (Steg, 2005).

With a theoretical point of departure in social practice theory, and an empirical base in qualitative interviews with adult Danes experiencing flight shame, this chapter sets out to explore how concerns about climate change might shape individual relationships to travel. Moreover, and of particular significance to this book, it draws on a social practice theoretical frame to examine how climate discourses are negotiated, transformed, and challenged when situated in the context of family travel. As such, this chapter aims to open the black box of family travel, to see how individual members contest and negotiate family travel practices, with a focus on how ethical positions in relation to travel and environmental consequences are taken up.

Methodological Note

This study is based on 13 in-depth research-interviews with Danish citizens, who claim to have experienced flight shame and have changed their travel patterns, so as to stop flying or at least fly less. Recruitment letters were shared in online fora for people engaged in different environmental initiatives, including debate communities, green student movements, and parent-solidarity networks for climate. Using a relatively open recruitment strategy, participants were sampled based on two criteria: one, that they had experienced feeling flight shame and two, that their concerns for the climate had somehow led them to change their travel behaviour. Aside from a short description of the emergence of the term, the recruitment letters contained no further definitions of what flight shame might be experienced as. Nor did it set any standards against which changes in travel behaviour could be defined. This allowed for a more inductive approach to understanding what flight shame entailed for the participants and to explore the different ways it shaped their travel practices. Since the initial open recruitment strategy led to responses from mainly young, well-educated women between the age of 20 and 35 years, this was supplemented by snowballing in an effort to ensure a more diverse sample in terms of gender and age (including participants who were parents). Except Emil, whose two children lived at home, the remaining participants were either young adults or parents whose children no longer lived at home. This meant that examining the negotiation of family travels was even more complex and insightful as families contended with adult children who had strong views on how and where to travel.

Findings

The following analysis has been divided into two main parts, exploring the concept of flight shame from different angles. In the first part, we explore the discursive and subjective components of flight shame as well as the consequences these have on the participants' experiences of air travel. Here, we draw on Kaufmann et al.'s (2004, 2017, 2018) notion of motility to examine how flight shame can be conceptualised from a mobility point of view, depending on the three modalities – access, capabilities, and desires – to appropriate mobility, and on Urry's (1990) idea of the *tourist gaze*, to explain how concerns for the climate pose challenges to the way travelling has previously been seen as a desirable practice. In the second part, we take on a social practice theoretical perspective to examine how flight shame is negotiated among family members in the context of family travel.

The Possibility-Desire Dilemma of Air Travel

The participants in this study were individuals, whose geographical, socio-economic as well as cultural situations allowed them to access a plethora of different forms of mobility and granted them the competences required to make use of this access. In fact, most of the participants had travelled extensively during their lives to date. Residing in more or less urban centres around Denmark – a country generally known for short distances, advanced transportation infrastructures, and high average income levels – most of the participants lived in relatively close proximity to airports, offering easily accessible spatial links to the entire world. Together with a relative abundance of economic means, and high educational attainment levels, far-distance travel occupied both an imaginable and a very real feature in these people's lives.

The past travel trajectories of the participants revealed a high level of ease related to being mobile. Mikkel, a 24-year-old student of engineering referred to himself as a 'merry traveller', and Caroline, a 27-year-old woman, who had recently started working in her first full-time job after completing her master's programme in societal studies, said that travelling had played a 'very big role' in her life:

> I have always travelled with my family at least once a year […] I have lived for half a year in Kenya and in South Korea and twice in Germany, so I have travelled a lot during the past few years, and it takes up a big part

of my identity. It has meant a lot for me, and it is something that I have done often, so I have been to many different places.

Caroline's account illustrates how long-distance travel represented a ubiquitous and easily accessible practice in her mind and life. She considered travelling an expected part of her life, to an extent where she felt it took up parts of her identity. Similarly, Marie explained how she used to travel by air extensively during her twenties, not only across Europe for leisure but also on a regular basis domestically in Denmark as a means of saving time when going to visit her parents in the northern part of the country. Emil, who worked with foreign development through exchange programmes in a large Danish NGO, flew several times a year for work, in addition to which he would usually take his two daughters on holiday to large European cities by plane. Even the two youngest participants in the study, Sofie and Nina (both aged 21 years), who had travelled by air the least compared to the other participants, had travelled by air during their sabbatical year as well as with their parents or on school trips while living at home. Characteristic for the participants was that they experienced few or no limitations to their potential or actual mobility as they possessed the necessary access and capabilities to be so (Kaufmann et al., 2018), hence having already accrued motility as a capital.

However, what characterised the accounts constructed by our participants was a desire to reduce or entirely abandon practices of air travel. A few of the participants, like Nikolaj and Marie, were hopeful about never having to travel by air again. They had both invested in alternative means of private transportation: Nikolaj had bought himself an electric car, which he aimed to use only when the weather conditions allowed him to drive on a hundred percent renewable energy; and Marie, who had been taking sailing classes and was currently saving up money to buy a sailboat, with which she hoped to travel the world one day. However, most of the participants were a little more reticent in their articulations of their future travel trajectories. They tended to emphasise the role of public transportation, like night trains through Europe, in order to reduce their air travel practices. Common for all the participants was that they felt caught in a dilemma between wanting to 'see the world', as Mikkel expressed it, and at the same time being genuinely concerned about the climate effects of air travel. This concern was perhaps most visible in expressions revealing the strong and negative symbolic meaning that airplanes had come to hold for most of the participants. Sara said,

> As soon as anyone even mentions that 'wouldn't it be exciting to travel there some day?', my mind is thinking CO2, CO2, CO2 […] When I am watching a movie, I cannot help but think, 'God there is a lot of CO2' – and [the same goes for] advertisements. I have this meditation-app with a blue sky and clouds on it – and then you suddenly see this airplane flying up there. And I am just like: No, no, no, it's an airplane!

Sara's account illustrates a clear association between the visual image of an airplane and the destructive potentials of the carbon it emits. Representative of the majority of accounts in this study, airplanes were understood as much in relation to their symbolic value and long-term effects (their harmful effect on the environment, their contribution to carbon-led climate change) as their instrumental functions (transporting people and things from A to B). Although to a varying degree, the image and mention of airplanes awakened negative and distressing emotions in the participants, recalling visions of climate catastrophe, or, as Sara later expressed it, 'the global doom'. Nikolaj evoked an imagery of airplanes as eruptive elements on an otherwise pristine blue sky, and Thea described flying as an 'absurd' and 'unnatural' practice, where so little thought was given to the amount of fossil fuel that went into making people 'hover above the clouds'. Laying the foundation for feelings of flight shame, it was the strength of these symbolic and affective responses to, and images of, air travel that fuelled the participants' reluctance to be mobile.

In a quantitatively informed study investigating individuals' and groups' aptitude for movement, Kaufmann et al. (2018, p. 6) propose a mobility typology in which those *reluctant to be mobile* can be said to have 'good access and skills but very little willingness to be mobile'. While their study attributed such reluctant attitudes to mostly demographic factors like gender, family structure, income, education, and identification with home country, these do not sufficiently explain the values, beliefs, and norms participants in this study expressed, and their rejection of the impetus to be mobile. Flight shame induced in the participants a kind of morally inclined reluctance to appropriate opportunities for air travel mobility. To understand these more discursive elements of flight shame, we draw instead on Urry's (1990) notion of the *tourist gaze*. He used the concept of the *tourist gaze* to explain how travel generates 'pleasurable experiences' of otherwise 'bizarre and idiosyncratic social practices' (Urry, 1990, p. 2). In the case of this study, though, a reverse process seems to have been taking place, in which the symbolic meaning of air travel has changed. This we might attribute to an idea of a *carbon gaze*. The *carbon gaze* should be understood as a way of seeing the world which orders it into objects that can be understood in accordance with their estimated impact on the climate. In this way, and continuing to build on Urry's argument, the development of a *carbon gaze* has worked to produce unpleasant experiences (of flight shame) out of the otherwise pleasurable social practice of travelling for the participants.

Where immobility for other groups of people presented in this book has been explained with reference to a disjunction between low access and capabilities, and high desires to be mobile, the dilemma for the participants in this study lies in the conflict between two opposing discourses, on the one hand incentivising them to be mobile and on the other condemning long-distance mobilities involving air travel. The fact that most of the participants had high accessibility and capabilities to be mobile seemed to only aggravate this conflict. It might then be suggested, that for these people, flight shame resulted from a predicament of having an excess of opportunity that was condemned as morally unjust by the *carbon gaze*.

The 'Awakening' of the Carbon Gaze

When asked about when and how they had become concerned with environmental and climate-related issues of consumption, none of the interviewees were able to pinpoint one specific moment or episode in their life. Rather, the urge to act in a more climate-friendly manner was described as occurring as part of a process of a more general change in their lives. Thea, for example, described it as a focus point that had 'come along' when she, in the beginning of her twenties, had gone to study in Copenhagen. Within a year she had moved into a communal living setting with other young people, met her boyfriend who volunteered for Green Peace, gotten her first student-relevant job in an NGO, and become vegetarian. She described it as 'having entered into something', and that 'something' made her 'realize that [she] was a political being'. Meanwhile, Victoria explained it as an 'awakening process' during her high school years, when she started participating in a left-wing youth political forum. Here, climate change was associated with social justice concerns, and made her realise that she 'had to take care of the environment'. To Thea this process made her realise that 'politics is more than simply party-politics and Danish politics; [that] you can also have political opinions, which you can express in different ways'. For many of the other participants, there was the feeling of political 'obligation' attached to this experience of awakening, alongside a sense that 'if [they] do not do it' – considering the amount of knowledge they have now – 'who will then do it'.

Common among these 'awakening' stories was that they involved a process of politicisation of everyday practices, that were suddenly being analysed and problematised from a climate-related framework of understanding. To most of the participants, this had led to a heightened focus on the individual moral responsibility of their own patterns of consumption. In Latour's (2014) words, such a change involves a transformation where *constative* statements, that is, objective statements about a natural phenomenon – in this case carbon dioxide emissions – come to be considered *performative*, in that they call on the listener to act – in this case by reducing one's carbon footprint. Applying this to our notion of the *carbon gaze*, we argue that while flight shame depends on scientific measurements to identify objects of climate concern and establish what *is*, these objective understandings also lead to the creation of moral judgements about what *ought* to be. The *carbon gaze* should, in Jasanoff's (2010) words, be understood as *co-productive* as it simultaneously produces the material world as well as confers normative authority of certain social practices over others.

What these accounts also illustrate is that processes of individual change were, more often than not, the result of changes in social circles or exposure. The kinds of institutions, networks, and situations that the participants found themselves in (be it jobs, educational institutions, or friendship groups) had, in other words, produced a sense of shared understanding of issues pertaining to climate change as well as the kinds of standards and practices considered necessary to combat such issues. Abstaining from flying was one such practice. This illustrates Warde's (2014, p. 295) point that changes in human behaviour and consumption are rarely the result of individual efforts alone, but often involve 'endogenous change in social circumstances'. The *carbon gaze* is, in other words, a socially organised disposition, negotiated and produced in and among social institutions and networks. In this sense, it can be understood as an expression of what Carfagna et al. (2014, p. 158) have identified as an *eco-habitus*, that is, a social disposition towards 'environmental awareness and sustainable principles', which, among other things, is associated with an aptitude for cosmopolitanism, idealism, and an affinity for the authentic (Holt, 1998). The last part is something we return to in the section that follows, when talking about the consequences of not abiding to the *carbon gaze*.

Dilemmas of the Self

When violated, this moral inclination to abstain from travelling by plane had negative implications on the participants' sense of Self. To most of the participants, air travel had come to be associated with different formulations of what Caroline explained as an 'inauthentic' practice. To Thea going flying evoked a fear of appearing 'hypocritical' as a result of 'not being able to live up to one's own values'. In Natascha's experience feeling shameful about flying was related to the thought of 'not doing enough about the climate'. Such accounts denote a tendency among the participants to relate private consumption of air travel to negative images of the Self. This self-referential component is clear in Marie's reflections about why she experiences flight shame:

> [...] for some it might seem out of proportions, but it is really a big deal for me, because I work with climate on a daily basis, so I feel as if [it is about] my integrity. I have tied some of my identity to this, after all, and it just never feels nice, or it is like compromising with one's own principles, right?

Similarly, Victoria explained how, to her, flying was linked to a concern that her actions might reflect badly onto her identity:

> I think it is a question of identity [...] I feel that one defines oneself immensely from one's actions, and if you make decisions, without any regard of both the close and the more far-off consequences, then I think, I would be extremely scared of being considered irresponsible, or yes maybe, mostly that thing about not being conscious. Not being deliberate. And not taking responsibility for my actions [...] It's a damn important personality trait, to take responsibility of one's actions and the words

one uses and the consequences one has on the world. And that is how I would like to be viewed. So if I decide to go to the Canary Islands, then I would think that I was making a decision that was so self-interested that it would reflect badly onto my character and my identity.

These accounts reveal that the experiences of flight shame are strongly connected to identity and the individual's ability to make one's actions 'reveal or produce the "true" self' (Arnould & Price, 2000, p. 140). As previously shown, such self-images had emerged through integrative processes which merged the participants' own beliefs and values with those of their immediate social contexts. Now, these contexts had also come to serve the indirect role of watchdogs. The participants' self-images were, thus, maintained through social interaction with friends, peers, or family, whose beliefs and/or actual reactions towards the participant's travel behaviour worked to reinforce experiences of (in)authenticity. Abstaining from flying might then be considered an *act of authentication*, that allows the participant to produce and maintain an appearance of oneself as a responsible and concerned consumer, actively invested in reducing one's personal footprint on the climate.

Despite the social nature of these identities, the sanctions for not abiding to one's true self were believed by the participants to be individual. To Emma, telling people that she had gone travelling by plane 'creates some conflict inside oneself', because going against one's principles reminds you 'that you are not complete after all; that you have double standards'. A few of the participants had consciously kept silent about their travel practices to colleagues or friends, to avoid feeling guilty about flying. Aside from reducing their travel activity, most of the participants had developed 'pro-environmental' habits: they ate vegan, vegetarian, or pescatarian food only, they bought organic products, and they had cut down their consumption of new garments drastically. In this sense, abstaining from flying was merely the newest change of behaviour among an already long list of acts that worked to produce an authentic self-image. This suggests that not flying was considered part of a parcel of leading a responsible and climate-friendly lifestyle.

Alongside its mental and emotional manifestations, flight shame was experienced physically by some of the participants. Marie, for example, referred to an episode of physical and emotional distress where she had burst into tears at a dinner party, because of the prospect of having to buy a plane ticket, to go see her boyfriend in France. Likewise, when asked if there was a physical component to her experience of flight shame, Victoria responded,

Most definitely! Yesyesyes. A lot. It is a shameful feeling; it is a defeat with regard to being principled, or just, or responsible. And it would be hard for me if people asked, "what will you do in your holidays", because then "urrghh … Well, I'm going to Switzerland, BUT …". I wouldn't want to say it.

Emma described feeling 'alien' when sitting in an airplane or airport:

> [...] compared to most other situations, I prefer not to make eye contact with people. I just plug music into the ears, and then try to get it over with. Music in the ears, eyes in a book and then get it over with. I mean I almost try to avoid having to deal with it.

Amalie explained how she sometimes had nightmares, in which she found herself sitting in an airplane. She would always wake up feeling distressed and frustrated. These accounts demonstrate how discourses around flight shame are not only rational axioms for behaviour but also emotionally and physically embedded inclinations that affect how the participants experience themselves and the world around them. The *carbon gaze* triggers a variety of thoughts, emotions, and sensations that are tied to constructions of self-identity. In this sense, the act of flying does not only lead to people have concerns about their impact on the environment but also has an effect on their identities, as the narratives of participants who have experienced flight shame demonstrate. It places people under a moral obligation to those objects it deems problematic, and consequently, the failure of the individuals to abide to these obligations causes feelings of inauthenticity.

While the first part of this analysis has explored the way(s) flight shame affects the individual's relationship to travel, in the second part of the chapter, we look at what happens when experiences of flight shame meet the often quite different expectations and social practices around family travel.

Negotiating Family Travel

From the perspective of social practice theory, leisure time air travel represents an interesting instance of consumption. Using Warde's (2014) distinction between *mundane* practices and *extraordinary* events, air travel can be said to fall somewhere in between the two. While family travel has come to constitute an expected, naturalised, and routinised part of many Northern middle-class family's annual life circle, it still represents an exceptional breaking of everyday life (Urry, 1992). Moreover, when situated, as family travel is, in the social context of family life, it becomes a *shared travel practice* (Barnes, 2001). This makes it subject to interactions that link conventions around *tacit* routines with *reflexive* negotiations between different members of the family, seeking to ensure a *coordination* of practices (Halkier, 2020).

According to the participants in this study, climate-related impacts of flying had never represented any larger cause for concern when considering their families' relationships to travel. As we shall see in this second part of the analysis, this placed the participants and their *carbon gazes* in an oppositional position to their families in every case. Their views contested existing routines and practices of family travel. Here, we look at how the participants experience this clash between *tacit* social practices of family travel and *reflexive* notions of the *carbon gaze*, and examine how negotiations over practices of family-making give shape to new practices around travel.

Relative Responsibilities

One recurrent theme in the interviews revolved around the extent to which the participants felt a sense of responsibility in relation to their families' travel practices. Interestingly, the majority of the participants had much lower – explicit as well as implicit – expectations of their family members, than of themselves, when it came to reducing carbon footprints. For Nikolaj, for example, it had been a conscious decision not to let his own concerns for the climate and environment affect his children's possibilities of travelling:

> [...] when we had our children, I thought, that it should not fall on them, that I have this thing [flight shame]. They should be allowed to have all the same possibilities, as everyone else. So I have also never tried to place it on them. They should have the same as everyone else. I have told them, that they should go and experience [...].

As the father in his family, Nikolaj's efforts to avoid projecting his own feelings of shame onto his children related to a desire to give his children all the same experiences as other children their age. Nikolaj, however, took a different stance from several other participants (in particular children of parents who flew extensively), who said that their reason for not expecting much of their families in terms of changing their travel practices grew from the evaluation that 'one cannot really do much about one's family', as Mikkel expressed it. This was related to the idea that efforts of trying would simply not bear fruit because families 'are as they are'. When wanting to share his own concerns about flying, Emil, who worked in an NGO, preferred directing the discussion to 'people who are my close friends, who are in jobs, that are similar to mine', and continued that he 'would not waste [his] efforts' talking to his family about matters regarding the climate, as they were 'in a different place' than himself. While Nikolaj's account revealed an inclination to value other virtues of family life, such as letting one's children explore the world on par with others, and that this should take precedence over concerns for the climate, Emil's reply indicated that differences in lifestyles among family members might simply appear too big a hurdle for him to feasibly to try and influence practices around travel.

To Marie, her decision to place family members under the moral obligation of the *carbon gaze* depended heavily on the type of relationship she had with a particular family member: 'I chose whom I shame', she admitted, smiling embarrassedly, before going on to explain:

> I think, it is easier for me to tell my mum: "I think this", and "this should be changed". And with my dad ... I think, I tend to smooth things out

more, I would rather have a conversation about all sorts of other stuff than this. [...] For example, I haven't asked him, if he compensates his air travel. I have asked that of my boyfriend. My dad is [in] this "safe-zone", so I don't demand anything of him [...] He is exempt from any kind of criticism. My mum isn't, so the good discussions lie with her [...] I think it would make me feel bad for him, is he was to have that discussion. And my sister is exempt as well. She is actually very aware about the climate and has introduced me to so much, but she's also free to fly as she pleases, I wouldn't even think about shaming her either.

Marie's story highlights the relational complexities of negotiating family travel practices. She adjusted her level of contestation according to an evaluation of whether the individual relationship to the family members could '(with)*stand*' the kind of discussion it would require. In the case of her sister, she assessed that because she had demonstrated concerns for the environment in other ways, she too was exempt from criticism.

Flight Shame versus Family Shame

Several informants spoke of the pressure they felt with regard to adhering to conventional travel practices within the family. Nina, a 22-year-old woman, spoke of an episode, where she had decided not to go on holiday with her family:

> Last year my parents went to Gambia on holiday, and I was invited to come along, and I chose to not go because I didn't feel like it. I mean, because I have told myself that I do not want to fly on charter holidays anymore. If I am going to fly, it should be because it has a real purpose, for example a semester abroad or that I fly somewhere for an educational year, or work. I declined my parents' offer to go on this holiday, where all my siblings joined, and it was extremely tough! It was extremely tough to tell them no, it was extremely tough not going, and it was extremely tough having to hear about it.

When asked how her family had reacted to her not joining the family's holiday, she replied,

> I don't know if I would say that they were sad about it, but they remain rather perplexed about it: "the plane flies anyway" and those kinds of things. But I think that they think – not all of them, but some of them – also think that it was brave of me to dare make that decision. Some of them told me so afterwards. The other [family members] have also said that they think it's a pity that because I care about the climate, is has to affect our family, so that we couldn't go on holiday together.

The idea that her own concerns for the climate had 'changed' her family in a negative way, speaks of the sacredness that routines around travel can have in a family (as illustrated in Chapter 4). It highlights the conflict between contesting notions of what the good life entails, where, from the perspective of the family, travel is considered a valuable and unique possibility of spending time together.

The social sanctions of contesting family travel practices included feelings of exclusion and segregation from the perceived unity of the family. Several of the participants explained how they struggled to escape the role within their families as 'that pious one' (Sara), who would involuntarily come to act as 'the other's bad conscience' (Nikolaj). Amalie said that she tried to sometimes spend less 'energy' trying to consume in a strictly 'green' manner, as it complicated her intentions of 'being a good sister'. This speaks of the challenges involved with going up against the routinised practices of family life. In Nina's case, the decision to turn down her family's offer to fly with them on holiday had left her in a dilemma. In her own words, Nina found herself in a kind of circular set of effects where concerns about her carbon footprint led her to feel 'flight shame', and feelings of not adhering to her family's expectations with regard to holidaying made her feel 'family shame'. Nina's account demonstrates how coalitions of incommensurable travel practices can cause emotional distress and a feeling of being 'trapped' with no real alternative for satisfying both the *carbon gaze* or expectations of how to perform family life. Despite her initial intention to refrain from flying, except in certain circumstances, Nina eventually agreed to fly with her family on holiday, when they invited her again, a year later:

> In fact, this year in the autumn break my family invited me on holiday again, where we were supposed to go to Cape Verde. They invited me in the spring, and I thought about it for a really, really long time, and then I actually ended up saying yes, simply because I think, it would have been too tough, if I wasn't going to join them again, and I wouldn't say that my family shamed me, but they made a big effort to tell me how sad they would be, if I chose to not go with them on holiday again [...] But as soon as I had told them that I would like to come, I regretted, because the other guilt was weighing so heavily on me.

A similar experience of navigating between one's own principles around flying and notions of 'the good family life' characterised Mikkel's interview. When his parents invited him and his sister to go with them to the Azores for Christmas, he initially told them that he did not want to join them. He thought it 'embarrassing' to tell his parents why he was not prepared to travel with them, and imagined his father ridiculing him for being 'naive' about the carbon impacts of his private consumption. 'When it finally dawned on him', Mikkel said, 'he just became SO upset, and it completely devastated me to see him that way. So that's when I realised that, OK, I can't cope with this. I can't cope with destroying the good family vibes?'. Like in Nina's case, Mikkel eventually surrendered to the wish of his father, as he felt that he would otherwise be responsible for creating family 'drama'. Curiously, and to Mikkel's own surprise, after their trip to the Azores, his father decided to spend his most recent holiday camping in Denmark, which, according to Mikkel, he had never done before, as Mikkel's parents' usual mode of transportation for holidays had always been to fly. This suggests that although the initial discussion between Mikkel and his parents did not lead to any immediate alignment of values, the discussion helped make visible the intersection between opposing modes of consumption. By openly contesting his family's travel culture, Mikkel explicated the tacit routines of travel to his father. This allowed for a kind of trade-off between different modes of consuming travel.

Compromising Travel Practices

Experiences like these might explain why some of the participants expressed some scope for negotiating travel practices within their families. As Mikkel later expressed it,

> 'I choose to believe and hope that it can change some of the culture: that I can get my parents to change, and then they talk to their friends and they realise how my parents have changed, because I have changed, and then it might be that my parents' friends also change, so that's the kind of chain-reaction I hope will happen. But, of course, I also really doubt whether it will happen'.

However, the overall impression gathered from the interviews was that while the participants expressed genuine interest in contesting conventional practices of travel, the complexities of family relations and the compelling role that travel played in processes of family-making made it difficult for them to abide to the imperatives of the *carbon gaze*.

This underscores Berthou's (2013) point that

> 'pro-environmental [in this case carbon neutral] practices should not be seen as one entity or well-defined set of isolated practices one can either choose to do or not to do, but rather as part of many other practices and fields in everyday life'.

p. 64

For Emma, a self-identified 'radical' in terms of concerns for the climate, this intersection between maintaining 'grounded' transportation practices and, on the other hand, seeing her family had come under serious strain since she moved to Northern Sweden. Yet her somewhat modest comment, 'I am a relatively social being, so it is very important for me, to be able to live a more or less normal social life', indicates the magnitude of the loss associated with living a life entirely guided by climate-friendly dogmas. Summing up the accounts of the participants, family travel, while an object of the *carbon gaze*, involved multiple negotiations over what it meant to be part of a family in a meaningful way, hereby rendering itself a complex social brico-lage (Bourdieu, 1977) of interrelated interests, ideas, and practices.

Conclusion

This chapter has explored the role of climate concerns on the potentials for, and practices of, air travel. The *carbon gaze* was identified as a useful con-ceptual tool for describing how climate-friendly dispositions among certain groups of Danes came to deem individual consumption of air travel prob-lematic. Situated in the social context of family travel, however, aspirations to reduce one's carbon footprint became part of a much wider network of practices that complicated the alignment of practices of travel across its mem-bers. The negotiations around travel practices in the families of the partici-pants demonstrates the role that air travel has come to play in family-making processes among modern Danish middle-class families. Although reflexive contestations around climate concerns hold the potential to make visible, and possibly change, tacit and routinised consumption patterns of family travel, the significance of those routines also complicates the link between value and action of those inclined to live more climate-friendly lifestyles.

Note

1 Like the British *Framework for Pro-environmental Behaviours* produced by the Department for Environment, Food and Rural Affairs (2008) or the endeavour of the United Nations Environment Program (UNEP) to make individuals 'Kick the CO_2 habit' (2008).

References

Alcock, I., White, M. P., Taylor, T., Coldwell, D. F., Gribble, M. O., Evans, K. L., Fleming, L. E. (2017). 'Green' on the ground but not in the air: Pro-environmental attitudes are related to household behaviours but not discretionary air travel. *Global Environmental Change, 42*, 136–147. doi:10.1016/j.gloenvcha.2016.11.005.

Arnould, E. J., & Price, L. L. (2000). Authenticating acts and authoritative perfor-mances: Questing for self and community. In S. Ratneshwar, D. G. Mick, & C. Huffman (Eds.), *The why of consumption: Contemporary perspectives on consumer mo-tives, goals, and desires* (pp. 140—162). Routledge.

Barnes, B. (2001). Practice as collective action. In T. Schatzki, K. Knorr Cetina, & E. von Savigny (Eds.), *The practice turn in contemporary theory* (pp. 17–28). Routledge.

Barr, S., Gilg, A., & Shaw, G. (2011). 'Helping people make better choices': Explor-ing the behaviour change agenda for environmental sustainability. *Applied Geogra-phy, 31*(2), 712–720. doi:10.1016/j.apgeog.2010.12.003

Berthou, S. K. G. (2013). The everyday challenges of pro-environmental practices. *The Journal of Transdisciplinary Environmental Studies, 12*(1), 1602–2297.

Bourdieu, P. (1977). *Outline of a theory of practice.* Cambridge University Press.

Carfagna, L. B., Dubois, E. A., Fitzmaurice, C., Ouimette, M. Y., Schor, J. B., Willis, M., & Laidley, T. (2014). An emerging eco-habitus: The reconfiguration of high cultural capital practices among ethical consumers. *Journal of Consumer Culture, 14*(2), 158–178. doi:10.1177/1469540514526227

Cohen, S. A., Hanna, P., & Gössling, S. (2018). The dark side of business travel: A media comments analysis. *Transportation Research Part D: Transport and Environment, 61*, 406–419. doi:10.1016/j.trd.2017.01.004

Dubois, G., & Ceron, J. P. (2006). Tourism/leisure greenhouse gas emissions forecasts for 2050: Factors for change in France. *Journal of Sustainable Tourism, 14*(2), 172–191.

Eickhout, B., & Taylor, K. (2016). Planes need to stop existing in a parallel universe when it comes to the climate fight. *The Guardian.* https://www.theguardian.com/environment/2016/sep/26/ planes-need-to-stop-existing-in-a-parallel-universe-when-itcomes-to-the-climate-fight (accessed on 12 December 2020).

Garnaut, R. (2011). *The Garnaut Review 2011: Australia in the global response to climate change.* Cambridge University Press.

Gössling, S. (2009). Carbon neutral destinations: A conceptual analysis. *Journal of Sustainable Tourism, 17*(1), 17–37. doi:10.1080/09669580802276018

Gössling, S., Humpe, A., & Bausch, T. (2020). Does 'flight shame' affect social norms? Changing perspectives on the desirability of air travel in Germany. *Journal of Cleaner Production, 266*(2020), 1–10. doi:10.1016/j.jclepro.2020.122015

Gössling, S., & Peeters, P. (2007). 'It does not harm the environment!' an analysis of Industry Discourses on tourism, air travel and the environment. *Journal of Sustainable Tourism, 15*(4), 402–417. doi:10.2167/jost672.0

Hache, E., & Latour, B. (2010). Morality or moralism? An exercise in sensitization. *Common Knowledge, 16*(2), 311–330. doi:10.1215/0961754x-2009-109

Halkier, B. (2020). Social interaction as key to understanding the intertwining of routinized and culturally contested consumption. *Cultural Sociology, 14*(4), 399–416. doi:10.1177/1749975520922454

Hares, A., Dickinson, J., & Wilkes, K. (2010). Climate change and the air travel decisions of UK tourists. *Journal of Transport Geography, 18*(3), 466–473. doi:10.1016/j.jtrangeo.2009.06.018

Hargreaves, T. (2011). Practice-ing behaviour change: Applying social practice theory to pro-environmental behaviour change. *Journal of Consumer Culture, 11*(1), 79–99. doi:10.1177/1469540510390500

Hibbert, J. F., Dickinson, J. E., & Curtin, S. (2013). Understanding the influence of interpersonal relationships on identity and tourism travel. *Anatolia, 24*(1), 30–39. doi:10.1080/13032917.2012.762313

Higham, J., Cohen, S. A., Cavaliere, C. T., Reis, A., & Finkler, W. (2016). Climate change, tourist air travel and radical emissions reduction. *Journal of Cleaner Production, 111*, 336–347. doi:10.1016/j.jclepro.2014.10.100

Higham, J., Cohen, S. A., Peeters, P., & Gössling, S. (2013). Psychological and behavioural approaches to understanding and governing sustainable mobility. *Journal of Sustainable Tourism, 21*(7), 949–967. doi:10.1080/09669582.2013.828733

Higham, J., Ellis, E., & Maclaurin, J. (2019). Tourist aviation emissions: A problem of collective action. *Journal of Travel Research, 58*(4), 535–548. doi:10.1177/0047287518769764

Holt, D. (1998). Does cultural capital structure American consumption? *Journal of Consumer Research, 25*(1), 1–25. doi:10.1086/209523

IPCC. (2013). *Climate change 2013: The physical science basis.* Contribution of Working Group I to the Fifth Assessment Report of the Intergovernmental Panel on Climate Change. Cambridge University Press.

Irwin, A. (2001). *Sociology and the environment: A critical introduction to society, nature and knowledge.* Polity Press.

Jacobsen, E., & Dulsrud, A. (2007). Will consumers save the world? The framing of political consumerism. *Journal of Agricultural and Environmental Ethics, 20*(5), 469–482. doi:10.1007/s10806-007-9043-z

Jasanoff, S. (2010). A new climate for society. *Theory, Culture and Society, 27*(2), 233–253.

Kaufmann, V., Bergman, M. M., & Joye, D. (2004). Motility: Mobility as capital. *International Journal of Urban and Regional Research, 28*(4), 745–756.

Kaufmann, V., Bergman, M. M., & Joye, D. (2017). Motility: Mobility as capital. *The City: Critical Essays in Human Geography, 28*(December), 337–348.

Kaufmann, V., Dubois, Y., & Ravalet, E. (2018). Measuring and typifying mobility using motility. *Applied Mobilities, 3*(2), 198–213.

Kollmuss, A., & Agyeman, J. (2002). Mind the gap: Why do people act environmentally and what are the barriers to pro-environmental behavior? *Environmental Education Research, 8*(3), 239–260. doi:10.1080/1350462022014540

Latour, B (2014). *War and peace in an age of ecological conflict.* Lecture prepared for the Peter Wall Institute Vancouver 23rd of September 2013 https://pwias.ubc.ca/videos/bruno-latour-war-and-peace-in-age-ecological-conflict-fall-2013-wall-exchange (accessed on 28 January 2021)

Mayor, K., & Tol, R. S. J. (2010). Scenarios of carbon dioxide emissions from aviation. *Global Environmental Change, 20*(1), 65–73. doi:10.1016/j.gloenvcha.2009.08.001

Peeters, P. M., & Dubois, G. (2010). Tourism travel under climate change mitigation constraints. *Journal of Transport Geography, 18*(3), 447–457. doi:10.1016/j.jtrangeo.2009.09.003

Shove, E. (2010). Beyond the ABC: Climate change policy and theories of social change. *Environment and Planning A: Economy and Space, 42*(6), 1273–1285. doi:10.1068/a42282

Shove, E., Pantzar, M., & Watson, M. (2012). *The dynamics of social practice: Everyday life and how it changes.* Sage.

Randles, S., & Mander, S. (2009). Practice(s) and ratchet(s): A sociological examination of frequent flying. In S. Gossling & P. Upham (Eds.), *Climate change and aviation: Issues, challenges and solutions* (pp. 245–271). Earthscan.

Steg, L. (2005). Car use: Lust and must. Instrumental, symbolic and affective motives for car use. *Transportation Research Part A: Policy and Practice, 39*(2–3), 147–162. doi:10.1016/j.tra.2004.07.001

Urry, J. (1990). *The tourist gaze: Travel and leisure in contemporary society (Theory, Culture & Society).* Sage Publications.

Urry, J. (1992). The tourist gaze "revisited". *American Behavioral Scientist, 36*(2), 172–186.

Urry, J., & Larsen, J. (2011). *The tourist gaze 3.0.* Sage Publications.

Warde, A. (2014). After taste: Culture, consumption and theories of practice. *Journal of Consumer Culture, 14*(3), 279–303. doi:10.1177/1469540514547828

Zalasiewicz, J. A. N., Williams, M., Steffen, W., & Crutzen, P. (2010). The new world of the anthropocene. *Environmental Science and Technology, 44*(7), 2228–2231.

10 Conclusion

Our Starting Point(s)

When Miri and Claire met in November 2016, and decided to work together on developing an empirical base for theorising more closely the lives of 'globally mobile families', we did not anticipate how central the concept of mobility would become for our work. Our engagement with how mobility shapes the identities, practices of education and family-making, and relations to the 'home nation' and school choice demonstrates how embedded it is in this group's everyday lives and the makings of futures (Yemini & Maxwell, 2018). It becomes embodied as a naturalised way of moving through social and physical space. It's an affective discursive trope that shapes how they evaluate opportunities for work, education, and travel. Globally mobile families, aside from the elites, are a group that appears to practice mobility almost without restriction, and in a continuous manner along various registers (relocating, travelling to visit family and friends, virtually connecting with colleagues and family around the world, and emplacing their futures within a transnational frame). Yet we argue that the imperative to think with mobility and aspire to be mobile shapes us all, far beyond only being the purview of the global middle classes.

Intrigued by the varied forms of mobility we found in our research on globally mobile families, we started to wonder about a kind of geographical mobility that many more families engage with, on a regular basis – travel abroad, usually for holidaying purposes. We were curious to understand what role short-term travel played in family lives – their educative purpose, how they might open up desires for mobility in the future, and whether or not they dismantled the orthodoxy of the nation-state as an organising structure (Beck, 2012) for their sense of self, future aspirations, and, critically, the social hierarchies within. So, with this thought, Miri decided to start interviewing families as she travelled to work conferences and on holiday. What we discovered through the narratives constructed along these journeys intrigued us. Social class – that is, being 'working class' or 'middle class', defined via occupational positions and level of education attainment – as a structuring mechanism mattered, but in ways we could not have predicted. We also discussed

DOI: 10.4324/9781003056430-10

our own approaches to holidaying with our family and realised for all our similarities (level of education, professional roles, approach to parenting), we also practised travel quite differently. Out of these surprising realisations, and the accompanying excitement that we were actually about to tackle a research topic that has not been studied much to date, emerged this book.

Through our mixed research design, we have been able to collect a set of materials which have begun to bring to life a multifaceted picture of family travel mobility today (Yemini & Maxwell, 2020), while also allowing us to integrate theoretical influences from Bourdieusian sociology, literature on parenting practices, concepts that consider people's engagements with the 'Other' – nature, peoples, cultures, and geographies – and scholarship that seeks to understand how we are approaching the catastrophic consequences of climate change. We could, of course, have sought to draw on different concepts, but our aim has been to examine the usefulness of these particular theories (ones we have drawn on in our other work in various ways) for elucidating family travel so that our book is not only a much-needed empirical base from which to think about short-term geographical mobility but also offers a relatively coherent frame for scholars to continue this work and add further dimensions to our thinking tools for this topic.

Critical to our journey in writing this book was first, inviting Katrine to join us in extending the focus of the book. She brought her wonderful writing skills to the project as well as fresh insight in various literatures and foci we could pursue. In the autumn of 2020 three undergraduate students also joined the team – Camilla Sofie Linander and Olivia Pauline Rud Mogensen, and Maluhs Haulund Christensen. They studied specific aspects of family travel as part of their undergraduate dissertations, and helped to shape Chapters 4 and 7, respectively. The second critical development in the book's journey was that most of the work took place during 2020 and early 2021, when COVID-19 managed to seep its way, in often devastating ways, into everyone's life. Many times, faced with the interminable sense that this pandemic and its closure of societies would never end, we wondered whether this book still had any relevance. Closure of borders, imposition of social distancing rules, staled commercial activity, severe interruption of education, overwhelmed medical systems and the rift between the 'haves' and 'have-nots' being cleaved open even more, would we ever return to 'normal' or would our family travel practices be altered forever? We were able to examine this specific question more closely in our book and were, dare we say, a little surprised how unaffected families appeared about the need to re-consider the meaning of, and possibilities for, travel 'after' COVID-19. We are eager to see, as the effects of the pandemic and the struggles to get it under control continue into the early part of 2021, whether these initial findings in our research will still be challenged. Some parts of the world may remain restricted for a while to come, other countries may decide to continue to manage their borders very tightly, and some people unwilling or unable to secure their

'COVID-19 vaccination passport' may never have the same freedom to holiday that they enjoyed before.

Alongside the seismic shock of COVID-19, another catastrophe continues to edge ever closer to an inevitable, far-reaching, attention-grabbing reality – climate change. Air travel, a popular and often necessary transportation mode for short-term holidaying, making accessible 'exotic' parts of the world, or upholding the possibility to return 'home' and/or see family, may never return to its previous glory. Online meetings and conferences are now the norm, hence the need for professional travel may be reduced forever, with organisations aware how much can be done, at much lower cost. The air travel industry may therefore have been changed forever due to the ravages of COVID-19, though the forces of capitalism probably mean even this industry will bounce back. However, the break from the everyday norm of accessible, cheap, and comfortable air travel might have made (many) more people realise – perhaps we do not need to, and perhaps we should not, be flying any longer. These questions will need to be carefully examined as we find our feet again in a COVID-19 world during the 2020s.

Theoretical Thinking about Family Travel

We started with Urry's (1990) classic work on the tourist gaze, where mobility is elevated into something out of the ordinary and special, and where meaning-making of places visited is negotiated across other's views of these places and also who we encounter while in these tourist spaces. We found that family travel held a hallowed place in the construction of family narratives, that the landscapes and cultures encountered were often awe-inspiring, and their time away was fun and exciting, but central to their stories was the fact of 'being together' as a family and creating long-lasting memories of adventure and reconnection (see, in particular, Chapter 4). Furthermore, we found that the value of a place was negotiated within social networks. This might be as a practice of distinction compared to others, such as the Levis' (in Chapter 5) or Liv's narrative (the young woman in Chapter 7). But this negotiation was most clearly illustrated during our discussion groups with young people (in Chapter 7), where the allure of the United States was constructed in one focus group, or the need to excuse and/or legitimate travel to only nearby neighbouring countries for holidays had to be in another. Critically, travel was found to be much more multifaceted as a family practice, than perhaps Urry's theorisation suggests, where 'tourism' was conceptualised more as an individualised experience. Meanwhile, maintaining social and family networks through 'visiting' as a form of holidaying (see the working-class respondents and our globally mobile families) or deep immersion in local cultures through long-term travelling (as with some of our 'nomadic' families in Chapter 8) appeared a more accurate description for these groups. Here, the reason and nature of family travel was substantively different, as was the engagement with the 'Other' this made possible. To understand this better, we drew on understandings of global mindedness (Andreotti et al., 2015) and different articulations of global citizenship education (Pashby et al., 2020) to get a deeper sense of these.

As sociologists, one of our primary concerns was to understand how family travel might be yet another mechanism through which social stratification was made possible as well as extended. At the same time, we remained eagerly open to the possibility that family travel might be able to disrupt these embedded cleavages. Here, we found Kaufmann et al. (2004, 2018) and a number of Kaufmann's other extensions to his theory of motility extremely useful. Kaufmann understands mobility as being shaped by social location, but as per the Bourdieusian influence, that mobility in turn can shape social structures. Linked to this dynamic is that the potential, aspiration, and/or desire to be mobile is also understood to have value, and, critically, that spatial mobility interacts with social mobility. This, in our view, offers specificity (breaking down mobility into 'access', 'competencies', and 'appropriation'), a broader frame of social organisation within which to place mobility, and finally, the necessary agency and creativity to see how mobility need not always be only a privileged practice, but could become one that is 'privileging' (Maxwell et al., 2018). In this way mobility is a capital that is already in play in shoring up social positions, a capital that can be further accumulated, but also a capital that can become activated if the necessary elements that make up mobility can be accrued.

Our analysis found that mobility is a classed practice, in that family travel is differentiated, at least to some extent, across social class lines (Chapter 4). Critically, all families realise travel's potential for accruing experiences and dispositions that will have value when negotiating their position within a social field (see Chapters 4, 5, 6, and 7). Yet the differences between more traditional social class lines are perhaps more unexpected. Thus, our working-class respondents with immigrant backgrounds, despite their usually quite limited economic resources and cultural integration (within Denmark and Israel – our two focus countries), actually travelled quite regularly and were able to use travel instrumentally to maintain transnational links, experience cultures through a 'visiting' lens, and actively inculcated in their children high aspirations for education and future employment through this 'opening up of the world' (see Chapter 5 and 7). In this way, mobility can be seen as a form of capital (i.e. motility) that was relatively easily accrued and seen as having a convertible value. Meanwhile, the global middle class families used mobility as a capital – to distinguish themselves, shape their identities, and ensure their children's resilience during future mobility (see Chapter 6).

A critical contribution these chapters have been able to make is the central role the nation-state played in family travel. For one, family travel was usually not seen as 'real' travel unless they left the borders of their home country (see Chapter 4). Second, our Israeli working-class and middle-class families actively used the nation-state in their work to accrue value from travel. This underlines the relevance of Kaufmann et al.'s (2004) concept of motility, but it also required us to draw on the concept of cosmopolitan nationalism (Maxwell et al., 2020) to fully understand the processes at work. The Israeli families we involved in our study used their engagement with the 'Other' through travel to actively shore up their children's national identities (as Israeli Jews) and commitment to planning their futures in Israel (see Chapter 5). This is a rather surprising finding in some ways, but one that also chimes with some of the comments made by young people (in Chapter 7) where comparing the Danish way of doing things with cultures experienced abroad, which, in turn, enabled them to understand better how Denmark was different (and often better).

No one to date has studied how family travel is a practice of concerted cultivation (Lareau, 2003). We found plenty of evidence to support this claim, and how it was occurring across differently socially located groups (see Chapter 5). This challenges the differentiation Lareau found between working-class and middle-class parents, as already suggested in Vincent and Maxwell (2016). However, travel as a practice of cultivation was not only instrumental, which is how it is usually conceived (Maxwell & Aggleton, 2013). Family travel was also found to be child-centred, keen to promote intimacy, and to strongly affect the practice of family-making. We argued in Chapter 4 that parents were, in fact, seeking to integrate travel abroad into a family habitus, so it became a naturalised and routinised practice that sought to connect emotions to places, and create memories that would be reproduced across generations. In fact, in Chapter 9, we see many stories of parents seeking to find ways to ensure their adult children continue to accompany them on holiday at least once a year, as they work to ensure the family stays connected.

Our analysis also highlighted the continuous negotiation of value that took place during discussions and positionings around travel. Young people in Chapter 7 could quickly feel excluded if they did not already possess motility, as Amalie demonstrated when she was effectively silenced during the group discussions, or in Oscar's rather defensive rejection of foreign travel as having any value. In these instances, we can understand family travel as not only having long-term implications for social position and social mobility (as the parents often understood it – as a cultivation practice whose effect would be reaped in the longer-term) but also as critical in shaping social hierarchies and identities of the self in the present. In Chapter 9, too, we saw an anxiety-ridden process of external and internal work done by our respondents who had experienced flight shame, as they tried to remain authentic to their beliefs, keep face in their close social and professional networks, but also still play the role of a 'good' family member. Our research is able to show how these negotiations are on-going in respondents' everyday lives – as they seek to change other people's views about air travel, seek to align the external and internal pressures to maintain a coherent sense of self, or navigate the regular discussions in young people's peer groups and in the classroom about 'Where are you going this holiday?' or 'What did you do in the break?'

The final theoretical element we added to our analytical frame was to consider different ways in which the anticipated and sought-after encounters with the 'Other' was conceptualised by our respondents. We wondered what kind of a cosmopolitanism was being expressed in these narratives and what consequences this might have for the types of relationships and knowledges international travel would yield. We drew on Andreotti et al.'s (2015) three distinctive dispositions around global mindedness and found that while most families were 'tourists' (as per Urry & Larsen, 2011), some displayed more 'empathy' or even the potential for 'visiting' engagements. The latter articulations are suggestive of a deeper, more respectful, equal relationship with peoples, places, and knowledge different to me/mine. It is the latter conceptualisations that are also sought after by Pashby et al. (2020) in their attempt to identify ways we could teach and engage people through global citizenship education to protect the environment, fight for social justice and practice de-coloniality. We argued that some of our globally mobile families and those introduced in Chapter 8 (the so-called nomads and air travel abstainers) move closer to an epistemological engagement with travel. These are interesting initial findings, which would need to be more deeply examined if we are to identify how family travel could play a critical in promoting a more just engagement with the world.

This book opens up more questions than it starts to answer. We must also acknowledge the Northern nature of the book in terms of who our respondents are and the theories that we used. Given the funding and resources available to us, this book has focussed on generating a range of data, from a number of perspectives in an attempt to start to uncover some important insights into mobility in the form of short-term family travel. Such a targeted data construction approach, with a flexible but overall integrated theoretical frame, has allowed us to find surprising and important practices within family travel that shape social class locatedness in the present and future, and re-consider questions of the 'global' and its immersion within the local. Critically, we shine a spotlight on family-making as central to travel-making and vice versa, ensuring that our scholarship will have relevance to colleagues working within sociology of the family and sociologists of education, as well as the much broader network of researchers interested in mobilities and processes of globalisation.

References

Andreotti, V., Biesta, G., & Ahenakew, C. (2015). Between the nation and the globe: Education for global mindedness in Finland. *Globalisation, Societies and Education*, *13*(2), 246–259.

Beck, U. (2012). Redefining the sociological project: The cosmopolitan challenge. *Sociology*, *46*(1), 7–12.

Kaufmann, V., Bergman, M. M., & Joye, D. (2004). Motility: mobility as capital. *International Journal of Urban and Regional Research*, *28*(4), 745–756.

Kaufmann, V., Dubois, Y., & Ravalet, E. (2018). Measuring and typifying mobility using motility. *Applied Mobilities, 3*(2), 198–213.

Lareau, A. (2003). *Unequal childhoods.* Berkeley.

Maxwell, C., & Aggleton, P. (2013). Becoming accomplished: Concerted cultivation among privately educated young women. *Pedagogy, Culture and Society, 21*(1), 75–93.

Maxwell, C., Deppe, U., Krüger, H.-H., & Helsper, W. (Eds.) (2018). *Elite education and internationalisation. From the early years into higher education.* Palgrave Macmillan.

Maxwell, C., Yemini, M., Engel, L., & Lee, M. (2020). Cosmopolitan nationalism in the cases of South Korea, Israel and the US. *British Journal of Sociology of Education, 41*(6), 845–858.

Pashby, K., da Costa, M., Stein, S., & Andreotti, V. (2020). A meta-review of typologies of global citizenship education. *Comparative Education, 56*(2), 144–164.

Vincent, C., & Maxwell, C. (2016). Parenting priorities and pressures: Furthering understanding of 'concerted cultivation'. *Discourse: Studies in the Cultural Politics of Education, 37*(2), 269–281.

Urry, J. (1990). *The tourist gaze: Travel and leisure in contemporary society.* Sage.

Urry, J., & Larsen, J. (2011). *The tourist gaze 3.0.* Sage.

Yemini, M., & Maxwell, C. (2018). De-coupling or remaining closely coupled to 'home': Educational strategies around identity-making and advantage of Israeli global middle-class families in London. *British Journal of Sociology of Education, 39*(7), 1030–1044.

Yemini, M., & Maxwell, C. (2020). The purpose of travel in the cultivation practices of differently positioned parental groups in Israel. *British Journal of Sociology of Education, 41*(1), 18–31.

Commentary

Final Reflections on Nurturing Mobilities

Paul Tarc

This intriguing book shines a much-needed light on practices of family travel and how they are intimately tied to processes of (familial) subject-making. With strategic intention, or more incidentally as a global middle class (GMC) 'habitus', experiences of family travel allow for the cultivation of certain dispositions, skills, and aesthetic sensibilities. These are both performative markers of cosmopolitanised social classes/groups as well as prospective forms of cosmopolitan/cultural capital that are expressed at home and abroad as 'expanded mindsets', deepened attachments to national or global identity, and are used for social class advantage under geometries of power and competition that stretch across political borders. Accordingly, the authors' extensive use of Kaufmann and colleagues' (2004) conception of 'motilities' is particularly fitting and revealing. The text also engages the conjunctures of travel and global citizenship education (GCE) and how ethical (reduced carbon footprint) and less touristic or consumptive forms of engaging 'the Other' might be possible. The ethical calculations that individuals make to avoid plane travel under 'flight shame' and how they inform their relations with other (less ecologically committed) family members are compelling and revealing of the complexities of awareness, commitments, and the politics of 'taking action'. The relations between travel and global citizenship education and how travel may be altered to reconfigure its educational effects are also compellingly conveyed. Indeed, this book made me think again about the complex relations between awareness and action (knowing and doing), between transnationalism/cosmopolitanisation and national identity-making, and international travel and a utopian desire for a global citizenship premised on equality and planetary health.

I am writing this 'commentary' from my perspective as a researcher of GCE and GMC; however, as I was reading through the well-theorised empirical accounts of this book, I couldn't help myself from newly reflecting on my own family's practice of travel/mobility and its relation to both our inherited social classed inclinations and our more recent social class-making over the last 20 years since our first child was born. My partner (also a university professor) and I have remarked that our children always seem somehow a bit more 'grown up' after a trip and I have reflected on how travel, like other significant life events, seems to mark out time or chunks of life history. However, this book presses me (and its readers) to consider more deeply how family travel shapes familial relations, self-making, (trans)national identity, and social class-making alongside the shaping of imagined (transnational) futures. From banal to exhilarating features, from new constraints/dependencies to new freedoms/interdependencies, and from being closer together to being further apart, *experiences of familial travel* – differentially engaged across geographical circuits and social groupings – *are indelibly a part of what makes modern subjects.* Maxwell, Yemini, and Bach illuminate and examine these elements in a small cross-section of families and youth from the Global North, setting a generative ground for further exploration of the many different kinds of families and family members engaging in practices of national and international travel.

The authors of this book cite Beck and his collaborator's challenge to 'methodological nationalism' and to privileging 'traditional class structures … as a mechanism for understanding social, economic and cultural relations and stratifications patterns today'; thus, these social theorists are implying that 'the notion of social classes' represents a 'zombie category', no longer fundamental as a way to explain individuals' and groups' more multifaceted and fluid affiliations and identifications under heightened cultural globalisation. I think this point is well taken. However, as dramatically illuminated across the chapters, social class positionings and re-positionings (as dynamic and always under production) are *highly constitutive* of the relational work of (family) travel and its uses. Indeed, parts of the book illustrate how (upper) middle classness is performed by travel. 'Flight shame,' for example, might not even register for the many kinds of families who have never or very rarely experienced air travel. Choosing not to fly for ecological reasons is a choice that likely emerges out of an elite habitus. The purposeful and strategic uses of travel by some families for cultural and educational benefits also illuminate a certain taken-for-grantedness of choice-making and autonomy made possible by having and expecting to have an 'expendable income'. Indeed, parts of the empirical accounts on 'middle class' families, at least from my social and geographical location, seem to be more reflective of upper-middle or urban-middle class families. Certainly, the anecdote of one participant's weighing in on the decision to fly or take one's 'motor home' is reflective of a very particular segment of 'global citizens'.

For this reason, I appreciate the authors' inclusion of working-class families in the Israeli context. While travel is important for these families, the qualities and uses of travel differ from their middle-class counterpoints. For working-class families, travel 'served to fulfil a complex combination of needs and wants', (p. 46), which suggests that travel for travel's sake was not common. The authors explain it thus:

> In our Israeli sample, another interesting opportunity for travel abroad was often made possible through the parents' workplaces – for example

when the supermarket department they worked in hit its sales target, or when a photographer was sent to China to purchase new equipment. These kinds of organised travel opportunities were more popular among working-class families, but were not mentioned by those families belonging to the Israeli middle-class segment.

Chapter 5 (p. 47)

In this finding we see how socio-economic class might set the conditions or 'broad parameters' for the travel itinerary. How working- and lower middle-class families travel might be more structured by the 'needs' of family visits and work, which then creates opportunities for other aspects and benefits of travel. But these other aspects (exploration, adventure, and visiting cultural sites) are likely more contingent on what unfolds and may not represent purposeful 'cultivation' practices.

At the same time, Maxwell, Yemini, and Bach assert that these working class, (largely) immigrant, families also are more likely to have experiences closer to the more normatively desirable form of travel as 'visiting', rather than 'touring'. This finding is very interesting, and it may be that linking travel to family and diaspora supports a more substantive form of cosmopolitanism. However, this finding might also have limits, or in some cases be more of an idealisation that limits social mobilities of minoritised peoples. While those with family dependencies are learning how 'real' people live in different places, less constrained upper middle-class families are free to see museums and events that accrue as cultural capital. Although dated, the case of my partner who visited her relatives near Calcutta on her few international travels from Canada to India as a child and teenager is a case in point. These family travels no doubt provided formative experiences on how others live in the world, but they also came with diasporic confusion, bouts of illness, family conflict, and little travel beyond the routing to the ancestral village.

For my own part, my family rarely travelled beyond day trips to local cities or some overnight camping. My rural middle-class background still shapes me even though as an adult and now as a professor I have done quite a bit of travel. When my wife and I were PhD students and then assistant professors we would organise trips with our growing family around the academic conferences in which we were participating. Travel for travel's sake has been less common for us, in part because of the expense of travelling with multiple children but also because we were not raised within an upper middle-class habitus. Our practices of family travel at the conferences were more functionally centred on navigating daily tasks over deliberately cultivating our children's cultural capital in the cities we were visiting. Now, with our occupational status and some cultural sensibilities of the upper middle-class as well as a more expendable income, our approach to family travel (and child-raising) has been still tethered to our respective rural and immigrant 'middle-class' backgrounds. Perhaps our children in their upper middle-class upbringing will be more purposeful with family travel. But the point here, for us, is that social class positioning is very tied into inter-generational family and diasporic relations and, indeed, is a very pertinent factor in our practices of family travel. While heightened mobilities and massifying travel across many social classes/groupings mark the 21st century, our own story suggests that these expanded opportunities of travel may still require a social class analysis.

The use of travel for Israeli families' 'cosmopolitan nationalism' was an intriguing finding. The authors highlighted the importance of Jewish identity and how this was cultivated through travel and engagements with 'the Other.' How these national identifications are developed or deepened through relations with others in distant places was well documented empirically and represents an important register for comparative research. As the authors note, the case of Israel does present unique features. Thus, it would also be interesting to compare middle- and working-class forms of family travel beyond Israel. Just for one example, I imagine working-class families in Canada whose 'travel' is circumscribed by road trips to proximate cities in Canada or nearby US states for hockey or soccer tournaments. For some families, 'extra cash' is funnelled into weekend hockey or soccer tournaments of a child who plays competitive sports. The whole family is dragged along into the intensive sport/hotel experience – coolers of food and drinks from home carried into hotel rooms, games of unsupervised chase with siblings in barely used stairways, rowdy team meals at local eateries, and the socialising of parents with the tense conversations about whose kid will get the scholarship. And if these families do fly internationally, what is cultivated in their travel practices? And how might a cosmopolitan nationalism emerge here as well?

Maxwell, Yemini, and Bach have engaged an understudied area that is an important component of subject-making processes in the global era. Conceptually, their studies show how travel re-arranges relationships that, in turn, shape (future) subjectivities and they raise questions of the limits and possibilities of cosmopolitan learning and ethical responsiveness under expanded cross-national mobilities. The empirical accounting is revealing and the analysis insightful. My commentary has only touched on a few elements within this provocative book. The individual chapters provide multiple entries for new angles of research on practices of (family) travel, GCE, and (global) class-making under heightened mobilities and more consolidated and consolidating diasporas.

Appendix
Methodological Choices and Considerations

Introduction

This book contains a variety of empirical data collected in the period from spring 2018 to fall of 2020. Situated mainly in the residential contexts of the three authors, the data was generated primarily, but not exclusively, in Denmark and Israel. The data collection process evolved across three phases. First, we interviewed families themselves 'on the move', going on a family holiday. We initially followed the principles of grounded theory (Charmaz, 2014), where the sample evolved in line with the analysis, allowing us to refine our theoretical sampling approach with the advance of our data analysis. This way, while the initial sampling was mainly convenient, meaning we interviewed families who were travelling and happened to be found in the same airport/heading to the same destination as ourselves, in the next steps of the interviews we aimed to follow the guidelines of theoretical sampling, by recruiting families with children of various ages, pursuing a variety of travel trajectories, and importantly, of different socio-economic statuses. The initial data provided a rich account of the families' travel routines, their perceptions of travel, and the possible advantages and disadvantages of such forms of travel for their children. This was the first bulk of data we collected and one journal article was published (Yemini & Maxwell, 2020) examining elements of our findings. It was also the writing of this paper which inspired us to write this book.

The original data then was supplemented with secondary data collected from travel blogs, family blogs and fora, and media reports (mainly featured newspaper articles). These sources of data were sought out in order to get an insight into how some families engage with alternative and unusual travel practices, not commonly found in the mainstream. Finally, a large bulk of the accounts presented here were collected through interviews, focus group discussions, and an open-ended survey, which focussed in on specific issues pertaining the role of travel, among parents and children, which had been shown to be significant in the first stages of the data collection, and critically relevant for the time during which the book was being written. Such issues included environmental concerns around flying by air, restrictions on movement during the corona pandemic, as well as children's and young people's understanding of travel. These accounts were collected to provide a broader set of more specific insights into topics related to the broad practice of family travel, and to expand the array of perspectives on these, including parents who were mobile for work, young people and young adults.

For most of the chapters in this book, the process of qualitative data analysis was not a distinct stage of the research; rather, it was constantly performed as a reflexive activity that informed further data collection, new angles for the analysis, yet further data collection, writing, and so forth (Charmaz, 2014). Thus, analysis did not constitute the last phase of the research process. It was an integral part of the research design and of the data collection (Coffey & Atkinson, 1996). Grounded theory (GT) provides rigorous yet flexible guidelines that begin with openly exploring and analysing inductive data and leads to developing a theory grounded in data (Strauss & Corbin, 1967). We chose to employ constructivist GT (Charmaz, 2014) in this book, taking into account the "relativity and subjectivity" (p. 14) of the subject and the research. GT was chosen first and foremost due to the current gap between theory and practice in the field of family travel, which suggested to us a need for an explorative and comprehensive understanding of the phenomenon. In the following we detail the three stages of data collection, and reflect on the empirical and methodological choices that form the basis of this book.

Interviewing While Travelling

This book is based on the use of an innovative method for gathering qualitative data through interviews, by interviewing families during the actual act of international travel, primarily by air. We approached families in airport terminals who were waiting to board, or during the actual flight journey and at airport playgrounds to discuss the practice of travel. We immersed ourselves in such moments, eliciting parental views, explanations, and rationalisations for travelling, as well as keeping an eye out for the kinds of encounters with different people, cultures, and forms of consumptions such mobility was understood to make possible. By approaching families at airports, and engaging with parents in discussions of travel, while they were in the midst of this endeavour, we gained a unique glimpse into parental mindsets, which in other settings might have been more carefully or reflectively narrated. Instead, we aimed to generate data while 'on the move' (Ross et al., 2009) between the families and ourselves. We argue that in mirroring the journeys of our participants we create a reflexive analysis of travel, where questions regarding purpose, anticipation, excitement, and hesitations were at the forefront for participants as they prepared to take off and land in new or familiar places.

This initial bulk of our engagement with the field, which has been published in Yemini and Maxwell (2020), led to the generation of 22 in-depth interviews with families (mostly mothers) in the process of travelling with their families. The shortest of the interviews lasted around 30 minutes, but a significant portion of them lasted for most of the flight, sometimes being 4–5 hours in length. The first 19 interviews were all performed with travellers on the following routes: Tel Aviv to Nuremberg, Nuremberg to Tel Aviv, Tel Aviv to London, London to Tel Aviv, Tel Aviv to Bangkok, Bangkok to Krabi, Bangkok to Tel Aviv, Nicosia to Tel Aviv. Most of the families were on one of these routes and were interviewed at the gate, but three families were interviewed in the airport terminal and followed a different route (Tel Aviv to Madrid, Tel Aviv to Helsinki, and Tel Aviv to Budapest). Families were travelling to or from vacation (10); while others were visiting family (9). We developed a detailed interview guide prior to the interviews which included topics related to families' routines during travel, including questions about destinations; frequency of travel; activities undertaken; planning; and the engagement with the children prior to, during, and after the travel. In addition, issues like parental aspirations for their children's futures were explored including families' routines at home, and extracurricular activities. Specifically, we asked how parents negotiated the differences they encountered in food, lifestyle, and level of development in the places they visited and would be visiting, as well as how they handled the difficulties of travelling with children. All names used in analysis are pseudonyms.

A convenience sampling method was employed to recruit families in the initial stages – that is, families who were travelling at the same time as the interviewing author, and who had children of school age. But in the later stages of the study, after 11 interviews, we were more selective in the kinds of families we approached – those accompanied by various numbers of children of differing ages, and those carrying smaller or larger luggage items, in order to increase the diversity within our initial sample. The final three families – 20, 21, and 22 – were recruited through personal links (though the families themselves had no direct connections to the authors) to specifically include families who held strong opinions about the act of travel – choosing not to due to environmental concerns (20, 21) and a mother (family 22) who struggled with the dilemma of travelling after the unexpected death of her husband. These families provided some 'outsider' insights on international travel and allowed us to reflect further on the themes that were initially developed via our analysis.

The final sample of this first block of data collection included mothers and fathers between the ages of 33 and 55. The participants' children ranged in age from two months to twenty-two years old, although all interviewed families had at least one school-age child. Participants included doctors, lawyers, academics, school teachers, social workers, cleaners, engineers, economists, sport coaches, those who were currently unemployed, students, and librarians. Most of our sample (15 families) could be broadly defined as middle class (based on education level attained and professional position), while some of the participants could be considered working class, as they held 'blue collar' positions (seven families). Those families who were professionally mobile for work, were categorised as members of the global middle classes (GMC) (four families). Due to our recent research on the GMC, and their increasing visibility in Israel, we felt a closer focus on the similarities and differences in their narratives would be of academic interest. Given that most of our recruited families in this initial stage ($N = 22$) were Israeli, it is important to note their immigration background, as it plays a significant role in shaping social class location in Israel. Most of them were second/third-generation Israeli (18 families) while four were first-generation immigrants (from former Soviet Union and France). In our analysis, therefore, we examined how migration histories and where their extended families were located might shape their orientation to travel. A second important note was that the interviewing author (Miri) is Israeli Jewish herself, first-generation immigrant from the former Soviet Union, while the collaborating author in this phase of the study (Claire) – involved in the design and analysis of the data – is white European. Both authors travel frequently for work, family vacations, and maintaining ties with extended family.

Interviews were not recorded, but extensive notes were taken during the interviews, which were reviewed and complemented with further details immediately after each interview. These data were discussed by the two authors in an on-going manner, in order to refine the interview guide, the sampling strategy and to start developing the thematic analysis. The analysis was conducted using the stages described by Thornberg and Charmaz (2014, pp. 156–158) and included initial (open) coding followed by focussed coding and theoretical coding. The notes and summaries were read repeatedly by the authors, looking for significant statements, sentences, or ideas that could be developed into "clusters of meaning" (Moustakas, 1994). These were searched within and across interviews to determine appropriate foci for analysis (as described by Curl et al., 2018). However, coding was not treated as a linear process similarly to the data collection; rather, so as to be sensitive to theoretical possibilities, we moved back and forth between the different phases of coding.

Secondary Data Collection: Blogs, Fora, and Online Data Collection

Charting the social and cultural landscape within which the families' experiences of travel occur and are shaped is a daunting task. The narratives and discourses presented and represented online are endless, many of them reproduced by an array of actors, many of which have (more or less hidden) commercial motivations. Over a 2.5-year period (2018–2020) we closely followed and continuously enlarged our collection of travel blogs, travel fora, and websites, all concerned with family travel. The data collected from these sources were mainly produced in English and Hebrew, and the initial searches included keywords such as family travel, air travel and families, expat families' groups in London and Berlin, and vacation and families. The searches were performed using Google search engine and specifically targeted Facebook and Instagram as these media are particularly widely used for sharing travel experiences (Fatanti & Suyadnya, 2015; Marder et al., 2019). While the data collected by these means was carefully gathered and analysed following principles of grounded theory, we also used these fora to recruit participants for our interviews, posting messages online and inviting them to consider participating.

As illustrated by Kozinets (2002), the first step of netnographic research is to identify qualified online communities that are appropriate to the research focus and research question. The online communities of interest should be relevant, active, substantial, data rich, and interactive. The five personal travel blogs used for analytical purposes in this book were selected by first searching the keywords *traveling/vacation with kids/children/baby*, and *blog* on google.com. The first 20 entries returned in each search were reviewed. The documents were analysed using the inductive method suggested by Strauss and Corbin (1997). An open coding procedure was conducted with line-by-line reading to identify the salient categories of information supported by the text. The constant comparative approach was used to saturate the categories. Three rounds of open coding were conducted to make sure that no drift occurred in the definition of codes or a shift in the meaning of the codes during the process of coding to ensure reliability (Gibbs, 2018). Axial coding was then conducted after the text was reduced to a small set of categories. During axial coding, the underlying uniformities in the original set of categories or their properties were discovered to formulate a smaller set of higher-level concepts. Lastly, selective coding was utilised when more abstract thematic patterns were identified, and the hierarchy was finalised. Operationally, the texts were read and decoded in the original language.

We believe that the constant comparison, multiple rounds of coding, a detailed description of the findings, and peer debriefing support the validity of the results. However, caution is advised regarding the potential bias in the interpretation resulting from the insiders' view. The growth of user-generated content (UGC) sites, through which tourists can discuss their holidays, provides an unprecedented opportunity to better understand tourists' experiences. According to the netnographic procedure outlined by Kozinets (2002), we identified qualified online communities relevant to the research topic and used them analytically to inquire into the ways travel is depicted and conceived by the authors and the responders.

(Re)opening the Field: Expanding the Topic and Perspectives on Travel

Once it had become clear that the initial analysis would eventually evolve into a book exploring a broader range of practices around and conflicts surrounding family travel, we branched out to include a wider variety of data into our analysis. This was also where Katrine joined the process of data collection and writing. In addition, three students studying an undergraduate degree in sociology from the University of Copenhagen (where Claire holds a position as a professor) became involved in the project, choosing to do their BA theses on related topics and in this way helped to generate valuable data and analytical drafts which formed the basis of two of the chapters of the book. A large bulk of accounts were generated through various qualitative methods, including interviews, focus group discussions, and open-ended surveys focussing in on specific issues pertaining to the role of travel, which had some way or the other shown itself to be promising in the first and second stages of data generation. Such issues included examining in much closer detail the travel habits of Danish families, environmental concerns around air travel and how these might influence travel practices, restrictions on movement during the corona pandemic as well as classed differences in young people's understanding of travel in a Danish elementary school context. Finally, an open-ended online survey was sent out to parents living across various countries in the Global North, distributed via peer networks. These accounts were collected in parallel streams to provide a broader scope of perceptions on topics related to the notion of family travel, and to expand the array of perspectives on the topic.

Survey-Based Interviews

The open-ended survey was administrated through the social networks of the first two authors and distributed via Facebook and WhatsApp as an anonymous Google docs form. We were able to approach a wide range of families who agreed to share with us their travel perceptions and trajectories, resulting in a total of 46 survey responses. The age of the informants ranged from 28 to 55 years, although the majority (67%) were between 40 and 49 years of age, and almost 70% of them were mothers. The largest group of informants (31%) lived in Europe, but the remaining respondents resided in various places around the world, including Asia (9), Australia (1), Central/South America (2), the Middle East (1), and North America (2). More than three thirds of the families had lived in another country other than their country of birth during their adult life. This was of particular significance to our previously mentioned interest in GMC families, and allowed us to further our analytical reflections concerning this group. Around one fourth of the informants could be considered upper middle class, holding positions as CEOs, managers, doctors, and engineers or working in consultancy or academia, while the majority worked in more typical middle-class jobs as teachers, journalists, and salespersons. Nine of them currently did not hold any paying job, and out of these one was retired.

Besides from a list of background questions, the survey consisted of seven open-ended questions, with multiple sub questions. The questions addressed topics around the families' travel habits, their motivations for travelling, the role of their children when travelling, environmental issues related to travel, as well as the potential benefits and challenges of family travel. As the data collection continued into the spring and summer of 2020, the effect of the COVID-19 pandemic on families' travel trajectories was examined as well.

In this regard, we were interested in parents' views on whether and how this recent development which had de facto halted travel altogether might alter future travel plans and trajectories. The open-ended nature of the survey allowed the informants to answer the questions in as much detail as they wished and generated quite lengthy personal accounts and perspectives. Since the survey was entered through an open link, the generation of accounts happened *asynchronous* to the process of posing questions (Brauns et al., 2017). While this obviously reduces central elements in building rapport with the informants, it also allows the informants to reflect on their past travel experiences and to build narratives around their experiences (Gibson, 2010).

Once the responses had been collected, they were filed into an Excel sheet. Descriptive statistics were generated from the informant's background information, thereby providing an initial overview of the population and building a comparative basis for the analysis of the different answers to the remaining survey questions. The responses to the seven questions were then transferred to NVivo, from where they were coded first through an initial coding process, and later through a thematic coding process based on key themes in the responses, and their comparative relevance to the remaining data that had been collected.

Interviews with Danish Parents on Holidaying and the Effect of Restrictions Due to COVID-19

Our initial conversations with Israeli parents around travel practices and aspirations, had made us interested in the habitual nature of family travel and the role of children in these. We were eager to explore how parents' accounts might differ in a national and cultural context different from of the Israeli one. We therefor sought out students who would be interested in doing a collaborative thesis project on the topic of family travel, and Camilla and Olivia took on the task of collecting data among Danish families.

The recruitment strategy for this study was based on two criteria, one being that the interview persons had to be currently residing in Denmark and the other being that they should have at least one child below the age of 25 years. The age criteria for the children was set based on an assumption that "children" above the age of 25 years were less likely to travel with their parents than children below that age. The extent and frequency of the families' travel patterns were not given any significance during the recruitment phase, as we were interested in learning about the varieties and differences in doing family travel. However, the initial round of recruitment, where potential informants were requested to contact us based on an online recruitment letter shared on Facebook, rendered very few responses. Our broad criteria of selection might have discouraged some people from contacting us, as they might not have felt that the letter addressed them specifically. Therefore, in the second round of recruitment, we used gatekeepers, found among friends and family, to get in contact with people who might be willing to share with us their travel stories. This resulted in the recruitment of nine informants of whom the majority were women (eight out of nine) and all were in heterosexual, monogamous relationships. Furthermore, all our informants were middle class based on their education level and employment positions. This was, in part, unsurprising, given the recruitment strategy we employed. Given the limited time the students had to do their thesis work, and the very real restrictions to recruitment and engagement due to COVID-19, the relative homogeneity of our sample was actually productive, in our view, as it meant we could focus on nuances across the sample without broader structuring factors (such as socio-economic position) explaining these differences away.

However, there are, of course, some reservations regarding the internal validity of the study (Thagaard, 2015, p. 204f), given we had mostly female respondents, all were middle class, in heterosexual relationships, and most of the informants lived in Zealand (eight out of nine), in close proximity to the capital of Denmark, and thus to transportation infrastructures like the central train station and Copenhagen airport. Still, we managed to recruit parents of a variety of ages, with different numbers of children, and – most importantly – with many different types of travel habits, which was of most importance to our initial focus of inquiry.

In our interviews we focussed on the subjective stories of parents regarding their experiences with and reflections about family travel. As the interviews were conducted during the second wave of corona-related restrictions of movement in Denmark (in the fall of 2020), the issue of the COVID-19 pandemic and its effects on families' travel trajectories was also addressed, in order to understand how such restrictions affected families' views on and desires to be mobile. The interviews took a semi-structured approach to the question of travel and its multiple purposes and meanings, allowing for flexible conversations where individual stories were encouraged and followed. Given the somewhat mundane nature of the interview topic, the researchers approached the interviewees with a *conscious nativity* (Kvale & Brinkmann, 2009), thereby encouraging the informants to elaborate their accounts and provide reflections on their decisions and perceptions to travel. As there exists minimal academic inquiries into the topic of family travel (Yemini & Maxwell 2020), we approached the topic from an explorative angle, investigating a range of different topics and themes related to family travel. Only later did we decide on the analytical approach to the chapter, and in this process, some of the topics that turned out not to be of relevance for the analysis were simply left unused. This included, for example, the informant's attitudes to the Danish authorities' handling of the corona pandemic, which was not included in the analysis.

Due to the COVID-19 situation, which had led to a lockdown of many parts of Danish society during the time of the fieldwork, the majority of interviews (eight out of nine) were conducted online, using Zoom or Skype for Business. In this way, we were able to avoid our informants feeling uncomfortable about breaching the authority's recommendations to keep social distance and only meet up with people in one's inner circle. The one physical interview conducted happened on the interviewee's own request. Although some aspects of building rapport with the informants might have been lost in the online interview format, the fact that the informants could conduct the interview from their home provided the informants with a comfortable and relaxed setting for talking about their experiences. This might have resulted in a more casual flow of the conversation, in which the informants could express themselves in a more relaxed manner, compared to if the interview had taken place in a completely new setting (Nehls et al., 2015, p. 146). It did, however, also result in a few interruptions from family members walking past.

A standardised transcription key was developed to ensure the study's *descriptive validity* (Maxwell, 1992, p. 285ff). The key ensured a standardised processing of the data, a correct representation of the informant's accounts, and a minimum of information loss. Breaks, gestures, and tone of voice were all included in the transcriptions, as these often contain valuable interpretive meanings that give nuance to the descriptive accounts of the informants. The interviews were listened to in several rounds. The coding process was a collaborative effort by the two students, who started out coding individually through an initial coding process, then met up to discuss which topics they had found and to explore new angles or interpretations of the data. Eventually, and after having consulted the theoretical basis of the analysis, eight thematic codes had emerged from the interviews. To ensure the anonymity of the informants, names, number of family members, age, and exact job descriptions were modified slightly in the final analysis. The age of the informants' children, however, remained the same, as these were considered significant for the analysis of the families' travel patterns and any changes to these that were described.

Interviews with Danes about Flight Shame and the Issues of Climate in Relation to Air Travel

Although environmental concerns of travel were not among the initial foci of this book (we were primarily interested in understanding how patterns of mobility could be conceptualised as a proxy for social stratification), we soon came to realise that the climate-related consequences of increasing global mobility has placed a serious question mark under our expected mode of travel (by air). We were therefore interested in understanding how discourses on climate change might have altered the way people and families relate to practices of travel. As part of her master's thesis dissertation in global development, Katrine thus decided to use as her starting point the generation of data on how experiences of *flight shame* have qualitative impacts on individuals' notion of travel.

Recruitment letters were shared in online fora for people engaged in different environmental initiatives, including debate communities, green student movements, and parent-solidarity networks for climate. Using a relatively open recruitment strategy, participants were sampled based on two criteria: one, that they had experienced feeling flight shame and the other, that their concerns for the climate had somehow led them to change their travel behaviour. However, aside from a short description of the emergence of the term, the recruitment letters contained no further definitions of what flight shame might be experienced as. Nor did it set any standards against which changes in travel behaviour could be defined. This allowed for a more inductive approach to understanding what shame entailed for the participants and to explore the different expressions it had on their travel practices. Since the initial open recruitment strategy led to responses from mainly young, well-educated women between the age of 20 and 35 years, this was supplemented by a snowballing technique, in which those participants already recruited were asked if they knew of anyone who would be interested in participating. This was done as part of an effort to ensure a more diverse sample in terms of gender, age, geographical location, and family status (including respondents who were parents).

Eventually 13 participants were recruited. Among these, the majority were women (ten), and almost half were students. The remaining participants held different jobs involved with city development, engineering, the civil service, project management, and consultancy. The group of participants were, in other words, highly educated and firmly situated in the upper middle class. One was retired after having worked in the education sector. The age of the participants ranged from 21 to 65 years with the majority of participants in their twenties (seven), and only two of them being parents. Except for one, none of the participants lived with their families, which enabled a new entry point for studying family travels, and how family travels are negotiated among adult family members living apart from each other. Characteristic for most of the participants was that they either studied or worked with issues pertaining to the environment or climate, or had dedicated a significant portion of their spare time to different activistic causes engaged with environmental and climate-related issues. The analysis thus reflects the experiences and perspectives of people who are in more or less daily contact with dilemmas around climate change.

Eight of the interviews were conducted in person either at participants' home or in a meeting room made available by the University of Copenhagen. The remaining five interviews were conducted online through Zoom. This was done on the request of the participants, either due to time constraints of meeting up in person or because of geographical distances (four of the participants lived outside of Copenhagen and travelling to meet them would have required significant resources both in terms of time and money). As these characteristics can be considered potentially influential on people's travel patterns and family lives, the use of online platforms for interviewing was instrumental in ensuring accounts from a wider variety of participants (Brauns et al., 2017), thus strengthening the validity of the data generated in relation to the goal of inquiry. The duration of the interviews lasted between one and two hours.

The interview guide took its point of departure in experiences with flight shame and was structured in a manner that initially invited the participants to share incidences and experiences related to their travel history, their entry into concerns with the climate, their experiences with flight shame, and incidences where this had caused conflicting situations with either family or friends. Later on, as the researcher had started to build rapport with the participants, they were asked to provide reflections concerning how flight shame had affected their relationship to travel, including family travel, and what aspirations they had for their futures in terms of being mobile. Across the interviews, special attention was given to the process of briefing and debriefing the informants prior to and after the interview. This was done for several reasons: first, because experiences related to shame potentially hold loaded and sensitive meanings and ramifications for the participants. As we have also described in the analysis, experiences of flight shame had many implications for the mental and social lives of several of the participants. It was therefore important for us to emphasise to the participants, that they had the right to not answer questions or to withdraw entirely from the interview at any given time, if they ever felt uncomfortable talking about their experiences. In the debriefing session following the interview, several of the participants expressed how sharing their thoughts and experiences during the interview had felt almost confessional, relieving them of the blame they often felt when thinking about air travel. All the participants were encouraged to contact the researcher, if they felt a need to discuss the interview session further, make elaborations related to their accounts, or correct statements they did not find had been adequately conveyed. The other reasons for emphasising the pre-briefing session were also related to the topic of shame, but had to do more with the reliability of the participants' accounts. To avoid having the participants unconsciously confirm any preconceived ideas about the moral opinions of the researcher, it was stressed that the point of the research study was not to infer any moral judgements on the behaviour of the participant, but to understand the subjective experiences of flight shame. It was also emphasised that there were no right and wrong answers to the questions asked by the researcher and that honest answers were appreciated. All the interviews were recorded and later transcribed verbatim.

The coding of the interviews was done using NVivo, and involved an initial open coding process, in which common themes and topics were identified and alternative takes on the data explored. Although several theoretical concepts were "tested" during the first steps of coding, the final theoretical framework was not provided until much later, leaving space for the empirical descriptions to guide the analytical process through its structuring stages. This broad and explorative approach allowed for the words and phrasings of participants' accounts to step into the foreground, thus giving the data analysis a more *emic* character (Maxwell, 1992, p. 289). Once the structure was established, a focussed coding process was executed, incorporating the-oretical themes. Names, ages, and professions have been changed slightly in the final analysis.

Focus Groups with Children on Travel

Responding to a call for research on mobility to include the voices and experiences of children, we were prompted to explore how young people perceive and aspire to be mobile through travel. We therefore conducted seven focus group discussions with young students attending the 9th grade at a public Danish secondary school. A total of 24 students participated in the discussions that were conducted in between classes and during lunch breaks. Considering that the participants were between 14 and 16 years of age, consent was given both by the school and their parents for young people to participate, but also by the young people themselves before the focus group commenced. The specific public school was chosen due to its diverse demographic base in terms of socio-economic and cultural backgrounds. Talking to children of varied backgrounds allowed us to explore how young people understand travel differently and in relative terms to their peers. Focussing on their past travel experiences and their future aspirations, the familiarity and diversity within the school context provided a useful setting for examining how travel ideals were negotiated and evaluated among children from various socio-economic and cultural backgrounds.

Our experience of the research was that we were able to generate fascinating insights into the practice of travel with young people, yet we had to be adaptable in the face of various challenges to creating a context in which such discussions could be facilitated. Many informants forgot about their interview appointments despite numerous reminders; students and teachers regularly interrupted the interviews as they walked into the classroom in which the interviews were being held; and the young people often changed their minds about whether they wished to be interviewed alone or within a group. The use of vignettes was productive for stimulating further exchanges between the participants, and also emphasised how views on, and desires around, travel appeared variable and to change during the course of the discussions, as meanings were negotiated and new insights formed through the group interactions. Extending the research with young people to include individual interviews and comparing constructions of travel between close friendship groups or more mixed cohorts of peers would also yield important additional insights.

Working with Language

The interviews in the first stage of our data generation were all conducted in Hebrew, the data compiled from online material were all in English or Hebrew, while the data in the third stage of data collection was either in

English (the survey) or Danish (interviews and focus groups). The data was transcribed in the original language and analysed in its original too. This was to ensure the authenticity of expressions was retained as much as possible, given different cultural inferences were important in the articulations of experiences and desires around family travel. Data was only translated into English when specific quotes were drawn on to illustrate an argument to be made in the book. In general, we tried to maintain the authentic voice of the parents and young people who participated in our studies, mindful of the potential danger of speaking for others or asserting our own opinions and perceptions over those expressed directly by the responders (Josselson, 2007).

Conclusion

The data generated for the purpose of this book represents a wide array of qualitative techniques for data collection and provides an illustration of the myriad ways social science research might approach the topic of travel. The most innovative aspects of our research design were, we argue, the interviews 'while on the move' and the focus group discussions with young people. The former allowed us to experience travel, at least for a short period of time, with our participants. The latter approach was critical in facilitating some sort of replication of how hierarchies within the social space of the school were being created and reproduced through the sharing of experiences and aspirations around travel.

Given the paucity of research on this topic, and the way social location, cultural context, and emerging political, public health, and environmental 'crises' come to impact family travel, we are very aware of the many gaps we have in the empirical material which informs the contributions made by the book. The need to examine the experiences of families living in several other contexts where social positionings are shaped by different structures is clear. The innovation that could be possible through researching families while 'on holiday', and comparing these to narratives constructed while 'at home' would also be fascinating.

The methodology employed here in both the data collection and analysis demanded from us to be continuously reflexive in relation to our own mobilities and choices around family travel. While the first part of data was collected before the COVID-19 pandemic, the second part was collected during this unprecedented change to 'travel as usual'. As the longer-term restrictions around COVID-19 remain uncertain, we wonder whether or not international travel by families will change more substantially than our initially analysis suggests. Furthermore, the analysis which emerged during our final substantive chapter (Chapter 9) is where we feel more focus is needed, as more consistent action is absolutely needed if the climate change crisis is not to engulf us entirely.

References

Brauns, V., Clarke, V., & Gray, D. (2017). Innovations in qualitative methods. In B. Gough (Ed.), *The Palgrave Handbook of Critical Social Psychology* (pp. 243–266). Palgrave.

Charmaz, K. (2014). *Constructing grounded theory.* Sage.

Coffey, A., & Atkinson, P. (1996). *Making sense of qualitative data: Complementary research strategies.* Sage.

Curl, H., Lareau, A., & Wu, T. (2018). Cultural conflict: The implications of changing dispositions among the upwardly mobile. *Sociological Forum, 33*(4), 877–899.

Fatanti, M. N., & Suyadnya, I. W. (2015). Beyond user gaze: How Instagram creates tourism destination brand?. *Procedia-Social and Behavioral Sciences, 211,* 1089–1095.

Gibbs, G. R. (2018). *Analyzing qualitative data* (Vol. 6). Sage.

Gibson, L. (2010). Using email interviews. *Realties' Toolkit, 9,* ESRC National Centre for Research Methods (NCRM).

Josselson, R. (2007). The ethical attitude in narrative research: Principles and practicalities. In D. J. Clandinin (Ed.), *Handbook of narrative inquiry: Mapping a methodology,* (pp. 537–566). Sage.

Kozinets, R. V. (2002). The field behind the screen: Using netnography for marketing research in online communities. *Journal of Marketing Research, 39*(1), 61–72.

Kvale, S., & Brinkmann, S. (2009). *Interviews: Learning the craft of qualitative research interviewing.* Sage.

Marder, B., Archer-Brown, C., Colliander, J., & Lambert, A. (2019). Vacation posts on Facebook: A model for incidental vicarious travel consumption. *Journal of Travel Research, 58*(6), 1014–1033.

Maxwell, J. (1992). *Understanding and validity in qualitative research.* Harvard Educational Review. Sage Publications, pp. 279–300.

Moustakas, C. (1994). *Phenomenological research methods.* London: Sage.

Nehls, K., Smith, B. D., & Schneider, H. A. (2015). *Video-conferencing interview in qualitative research. Jew-Hai, Shalin: Enhancing qualitative and mixed methods research with technology.* Information Science Reference, pp. 140–157.

Ross, N. J., Renold, E., Holland, S., & Hillman, A. (2009). Moving stories: Using mobile methods to explore the everyday lives of young people in public care. *Qualitative Research, 9*(5), 605–623.

Strauss, A., & Corbin, J. (1967). *Discovery of grounded theory.* Hawthorne, NY: Aldine Publishing Company.

Strauss, A., & Corbin, J. M. (1997). *Grounded theory in practice.* Sage.

Thagaard, T. (2015). *Systematikk og innlevelse - En Inn føring i kvalitativ metode.* Fagbokforlaget.

Thornberg, R., & Charmaz, K. (2014). Grounded theory and theoretical coding. In U. Flick (Ed.), *The SAGE handbook of qualitative data analysis* (pp. 153–169). Sage.

Yemini, M., & Maxwell, C. (2020). The purpose of travel in the cultivation practices of differently positioned parental groups in Israel. *British Journal of Sociology of Education, 41*(1), 18–31.

Index

For Product Safety Concerns and Information please contact our EU
representative GPSR@taylorandfrancis.com
Taylor & Francis Verlag GmbH, Kaufingerstraße 24, 80331 München, Germany

www.ingramcontent.com/pod-product-compliance
Lightning Source LLC
Chambersburg PA
CBHW060318220326
41598CB00027B/4357

* 9 7 8 1 0 3 2 1 1 4 8 1 1 *